PRAISE FOR LUKE MA

This is a fast-paced, jaw-dropping tour of the ocean's most misunderstood predators. Funny, fearless, and fiercely factual—perfect for curious readers of any age.

— DR. ISLA WAVE, OCEAN WONDERS INSTITUTE

From glow-in-the-dark sharks to river-dwelling marvels, it turns wild trivia into solid science you can trust.

— CAPTAIN REED TIDE, EXPLORER AND SHARK EDUCATOR

A treasure chest of trivia that teaches as it entertains—my students begged for more after page one.

— ALEX MARIN, MIDDLE SCHOOL SCIENCE TEACHER

THE ULTIMATE BOOK OF CRAZY SHARK FACTS

THE ULTIMATE BOOK OF CRAZY SHARK FACTS

THE ULTIMATE BOOK OF CRAZY SHARK FACTS

A COMPLETE COLLECTION OF OVER 1500 UNIQUE
FACTS ABOUT SHARKS

THE GREATEST CRAZY FACTS SERIES
BOOK ONE

LUKE MARSH

To the ocean, our shared home, and to every curious mind who has ever wondered what lurks beneath the surface. To the sharks —ancient, misunderstood, and endlessly fascinating—may you always swim free, and may the truth of your world be shared with humility and wonder. This book is for the scientists who map the deep, the divers who witness it, the students who ask why, and the readers who choose wonder over fear. May these pages illuminate, educate, and inspire a future in which awe for the sea goes hand in hand with protecting it.

The sea, once it casts its spell, holds one in its net of wonder forever.

— JACQUES COUSTEAU

CONTENTS

Introduction   xiii

1. Meet the Sharks: Not All "Jaws" Are Alike   1
2. Built to Bite: Teeth, Jaws, and Never-Ending
   Replacements   9
3. Skin Like Sandpaper: Scales, Speed, and
   Secret Armor   17
4. Super Senses: Smell, Hearing, and Detecting
   a Drop of Drama   21
5. Electric Powers: The Sixth Sense That Finds
   Hidden Prey   31
6. Glow-in-the-Dark, See-in-the-Dark: Deep-
   Sea Shark Weirdness   41
7. Sharks That Walk, Wiggle, and Pull Off
   Weird Moves   51
8. Tiny Terrors to Ocean Giants: Record-
   Setting Sizes   61
9. Strange Bodies Department: Heads, Fins,
   and Other Bizarre Parts   71
10. The Menu: What Sharks Eat (It's Not Just
    "Anything")   79
11. Feeding Frenzy Facts: From Calm Cruising
    to Chaos   87
12. Baby Sharks! Eggs, Pups, and Reproduction
    That Gets Wild   97
13. Growing Up Shark: Lifespans, Maturity, and
    Ancient Survivors   105
14. Sharks on the Map: Where They Live and
    How They Travel   115
15. Freshwater and River Sharks: When
    Saltwater Rules Don't Apply   127
16. Teamwork, Attitude, and Social Lives: Shark
    Behavior Up Close   137

17. Myth-Busters: What Movies and Legends
    Got Totally Wrong   147
18. Humans vs. Sharks: Conservation, Threats,
    and Comebacks   149
19. Shark Science and Tech: How We Study
    These Ocean Icons   159

INTRODUCTION

Welcome to the wild, toothy world beneath the waves. The Ultimate Book of Crazy Shark Facts is a fast-paced tour through the ocean's most misunderstood predators, packed with jaw-dropping truths, surprising science, and unbelievable stories. Meet sharks that glow in the dark, those that can "walk" along the seafloor, and ancient species older than dinosaurs—plus record-breaking bites and bizarre body parts that defy expectations.

You'll discover how sharks sense electricity in the water, why some can survive upriver, and how fast they can strike when prey appears. We pull back the curtain on feeding frenzies and reveal what really happens, separating cinematic myth from real biology, with humor and awe to keep curiosity buzzing.

With topics spanning evolution, anatomy, behavior, and conservation, this book offers bite-sized trivia for curious kids, trivia fans, and ocean lovers of any age. Whether you read straight through or flip to a random page for an instant mind-blown moment, this guide makes sharks feel unforgettable—and you'll want to share every fact that surprises you.

MEET THE SHARKS: NOT ALL "JAWS" ARE ALIKE

Sharks aren't all teeth and terror; they're a dazzlingly diverse group with a toolbox of shapes, sizes, and lifestyles. In this chapter, we map the big families, spotlight oddball species, and give you quick ID clues that separate sharks from other fish.

1 - There are roughly 500 living shark species.

2 - Sharks are cartilaginous fish, meaning their skeletons are made of cartilage rather than bone.

3 - Sharks have five to seven gill slits on the sides of their heads, depending on species.

4 - The seven orders of sharks are Hexanchiformes, Heterodontiformes, Orectolobiformes, Carcharhiniformes, Lamniformes, Squaliformes, and Pristiophoriformes.

5 - Hexanchiformes include the sixgill and frilled sharks, among the most primitive living sharks.

6 - Heterodontiformes include the bullhead sharks, known for their blunt heads and unique teeth.

7 - Orectolobiformes are the carpet sharks, including nurse sharks and wobbegongs.

8 - Carcharhiniformes is the largest order and includes many familiar sharks like tiger sharks and hammerheads.

9 - Lamniformes are the mackerel sharks, including great whites, makos, and threshers.

10 - Squaliformes include the dogfishes, a diverse group often found in cooler waters.

11 - Pristiophoriformes include the saw sharks, named for their long, toothed rostrums.

12 - Hammerhead sharks have a distinctive cephalofoil that broadens their sensory field.

13 - The cephalofoil also provides improved binocular vision for hunting.

14 - The sixgill shark has six gill slits, a feature distinguishing it from most other sharks.

15 - Frilled sharks have an eel-like body adapted to deep-sea life.

16 - Lanternsharks produce light along their bodies via photophores.

17 - Bioluminescence in lanternsharks helps camouflage in the deep and communicate with kin.

18 - Goblin sharks possess jaws that can protrude outward, snapping up prey quickly.

19 - Goblin sharks are deep-water specialists with long, slender snouts.

20 - The Greenland shark can live for centuries and grows very slowly.

21 - Whale sharks are the largest living fish and feed on plankton by filtering water.

22 - Basking sharks are giant filter feeders with enormous mouths and gills.

23 - Nurse sharks are bottom-dwellers that commonly rest on the sea floor.

24 - The sand tiger shark has a mouth full of visible teeth even when its jaws are closed.

25 - The cookiecutter shark leaves circular wounds on larger animals.

26 - Thresher sharks use their unusually long tails to stun prey.

27 - The great white sharks are apex predators with powerful bites to grab and slice prey.

28 - Great white teeth are triangular, serrated, and continually replaced.

29 - Shortfin makos are among the ocean's fastest swimmers.

30 - Longfin makos are close relatives with similar speed and efficiency.

31 - Hammerhead sharks often form schools during migrations and feeding events.

32 - The eyes of hammerheads are placed on the edges of the cephalofoil for a broad field of view.

33 - Whale sharks feed by filtering plankton as they swim with mouths wide open.

34 - The basking shark also filters plankton with a huge mouth and long gill arches.

35 - Nurse sharks feed mainly on invertebrates and fish found on the bottom.

36 - Leopard sharks are easily recognized by their spotted markings along the California coast.

37 - Requiem sharks include several coastal species like the silky shark and blacktip reef shark.

38 - White-tip reef sharks hunt at night around coral reefs and then bask in the daylight.

39 - The sixgill shark is a deep-water species with a large body.

40 - Many sharks give birth to multiple pups at once, while others have only a few.

41 - Some sharks lay eggs in protective cases called mermaid's purses.

42 - Other sharks are ovoviviparous, with pups developing inside eggs that hatch inside the mother.

43 - Some sharks are viviparous, giving birth to live young with placental connections.

44 - Sharks continually replace their teeth throughout life.

45 - Teeth shapes vary by diet: serrated triangles for slicing meat, needle-like shapes for gripping fish, or flat crushing teeth for hard-shelled prey.

46 - Sharks have dermal denticles on their skin, making it feel like sandpaper and reducing drag.

47 - The streamlined bodies of sharks reduce drag and enable fast, efficient swimming.

48 - Sharks lack a true swim bladder; buoyancy comes from heavy livers and cartilage.

49 - The ampullae of Lorenzini are electroreceptors that help sharks detect the electrical fields of prey.

50 - The lateral line system detects water movements and vibrations.

51 - Many sharks have excellent night vision and can see in low light.

52 - Some sharks can tolerate brackish water and even venture into rivers, like bull sharks.

53 - The growth rates and age of maturity vary widely, with some species maturing late.

54 - Some sharks mate through internal fertilization and have internal development of embryos.

55 - Some sharks have complex social structures, including transient groups and mating displays.

56 - Sharks can migrate enormous distances, crossing entire ocean basins.

57 - The feeding ecology of sharks ranges from planktivores to apex predators.

58 - Ocean currents and magnetic fields guide many shark migrations.

59 - The countershading on many sharks helps conceal them in open water.

60 - Some sharks can swim at low speeds to conserve energy, especially when hunting.

61 - The teeth of small sharks are small, while large sharks possess larger teeth.

62 - The jaws of many deep-sea sharks are highly adaptable to prey availability.

63 - Saw sharks use their rostrum to sense and slash prey.

64 - Frilled sharks are pale in color and have a slow, deliberate hunting style.

65 - The deep-sea goblin shark uses its jaw to prey on other fish.

66 - The diversity of sharks includes both active predators and gentle filter feeders.

67 - Some shark species are particularly vulnerable to overfishing, making conservation important.

68 - Sharks contribute to healthy coral reefs by controlling prey populations.

69 - The film Jaws is often misinterpreted; real shark behavior is more nuanced and varies by species.

70 - The fossil record shows sharks have existed for hundreds of millions of years.

71 - The Megalodon is an extinct giant relative of the modern great white.

72 - The basking shark has a slower life history than other large sharks.

73 - The tiger shark has a broad, adaptable diet including a wide range of prey.

74 - Leopard sharks have distinctive spotted coloration that helps with camouflage along shorelines.

75 - Nurse sharks have mouths lined with small teeth that help crush prey.

76 - Wobbegongs rely on camouflage to ambush prey on the reef floor.

77 - Coral reefs host a rich diversity of small sharks adapted to complex habitats.

78 - Open-ocean pelagic sharks roam vast distances in search of food and breeding grounds.

79 - Shark diversity is a key indicator of ocean health and biodiversity.

80 - Understanding shark diversity helps protect ecosystems and informs conservation strategies.

81 - Whether you love tiny lanternsharks or giant whale sharks, sharks showcase remarkable specialization and resilience.

2

BUILT TO BITE: TEETH, JAWS, AND NEVER-ENDING REPLACEMENTS

From conveyor-belt teeth to never-ending replacements, this chapter dives into sharks' toothy world. Discover how jaws are built for bite after bite, and why losing thousands of teeth over a lifetime is just part of a predator's design.

82 - Shark teeth are not rooted in sockets like human teeth; they are anchored by ligaments to the jaw.

83 - Each shark jaw contains multiple rows of teeth that form a conveyor belt system.

84 - When a tooth is lost, a replacement tooth moves forward to replace it.

85 - A shark can carry many teeth at once, with dozens to hundreds in the mouth at any time.

86 - Tooth shapes vary by species and diet: some teeth are sharp and pointy, others are broad and flat.

87 - Serrated edges on teeth help slice through flesh.

88 - Pointed teeth are great for gripping slippery fish.

89 - Flat, crushing teeth are useful for eating hard-shelled prey like mollusks.

90 - Replacement rates differ by species and relate to how aggressively a shark feeds.

91 - Replacement teeth form in a growth zone behind the active tooth and move into place as needed.

92 - The enamel on shark teeth is very hard, helping them resist wear.

93 - Juvenile sharks may have different tooth shapes than adults as their diet changes.

94 - Teeth wear down and are routinely replaced to keep the bite effective.

95 - The term polyphyodont describes the endless tooth replacement in sharks.

96 - Some sharks have several rows of teeth that can replace each other.

97 - Ancient sharks such as Helicoprion had unusual tooth arrangements, such as a spiral tooth whorl.

98 - Megalodon teeth are among the largest known shark teeth and indicate a huge bite.

99 - Fossil teeth reveal how ancient sharks adapted to different prey and environments.

100 - Most sharks bite and tear their prey rather than chew it.

101 - Shark jaws are made of cartilage, not bone, making them light and flexible.

102 - Teeth are anchored in the jaw by connective tissue, not rooted sockets.

103 - New teeth push forward as old teeth fall out, keeping the bite stocked.

104 - Some sharks have curved teeth that help prevent prey from escaping.

105 - Teeth are arranged in patterns that help with gripping and tearing.

106 - The arrangement of teeth can indicate a shark's typical prey.

107 - Some sharks have larger front teeth and smaller back teeth for efficient capture and processing.

108 - The crown is the part of the tooth that does the biting; the root anchors the tooth.

109 - The enamel is the outer, toughest layer of the tooth crown.

110 - Dentin lies beneath the enamel, adding strength to the tooth structure.

111 - Cookiecutter sharks have small, sharp teeth that can gouge circular wounds.

112 - The jaw movement of sharks helps them grip and pull prey tighter.

113 - The presence of multiple tooth types in a mouth supports handling different prey textures.

114 - The teeth of a large predator can be credited with its ability to capture and hold onto large prey.

115 - The crown's curvature and serration pattern provide clues about a shark's hunting style.

116 - The enamel's microstructure helps reduce wear and tear during feeding.

117 - Ancient sharks show a variety of tooth shapes not seen in living species.

118 - Fossil teeth reveal ancient marine ecosystems and predator-prey relationships.

119 - The belt-like arrangement of teeth allows one tooth to replace another mid-bite.

120 - The teeth are constantly renewed, reducing the risk of long-term tooth loss.

121 - The presence of many teeth can be a coping mechanism during fights with prey.

122 - The dental system is a key adaptation for predation and feeding efficiency.

123 - Some sharks have teeth that are particularly tough for cracking hard prey.

124 - The tooth replacement cycle is an essential feature of shark ecology.

125 - Teeth can fall out while the jaw continues to bite, thanks to the backup teeth.

126 - The circular cookiecutter bite exemplifies specialized dental adaptation in predation.

127 - Sharks can replace teeth very quickly, maintaining a full bite during hunting.

128 - The dental system supports a variety of feeding strategies across different shark lineages.

129 - The tooth replacement belt allows continuous predation even when some teeth are damaged.

130 - The crowns of teeth in different sharks can have different edge patterns depending on prey type.

131 - The enamel margin can show wear patterns that reveal feeding history.

132 - Fossil records provide insights into how ancient sharks hunted and where they lived.

133 - A single bite can engage several teeth as the jaws close together.

134 - Tooth wear patterns can reveal whether a predator hunted mobile or stationary prey.

135 - The replacement tooth behind the active one is primed to slide into place when needed.

136 - The front teeth often do the initial gripping, while back teeth assist in processing.

137 - The enamel's durability helps teeth resist chipping during bone-cracking bites.

138 - Some sharks have teeth that are especially well-suited for piercing or gripping rather than cutting.

139 - The arrangement of teeth across the jaw helps maintain bite strength when prey fights back.

140 - Tooth shapes are a key cue in identifying shark species from fossils.

141 - Shark teeth can tell us about ancient ocean temperatures and prey availability.

142 - Megalodon's teeth reflect a huge predatory niche and provide clues about its prey.

143 - Fossil teeth show a long history of dental innovation among sharks.

144 - Cookiecutter bites demonstrate how tooth design can be specialized for precise tissue removal.

145 - The continuous renewal of teeth is a major factor in a shark's success as a top predator.

146 - Tooth shapes continue to diversify in response to changing oceans and prey communities.

147 - Sharks rely on a variety of tooth shapes to tackle a broad range of hunting scenarios.

148 - Tooth replacement is a robust safety net against feeding disruption due to tooth loss.

149 - The crown-to-root ratio varies across different tooth types within a shark's mouth.

150 - Teeth can be used as a fossil record of shark evolution and ecological shifts.

151 - The front-most teeth are often the most worn due to their role in initial prey capture.

152 - Sharks' tooth replacement systems are among the most successful biological solutions for predation ever evolved.

153 - Tooth replacement is integrated with growth and development throughout a shark's life.

154 - The diversity of crown shapes among sharks is a testament to ecological diversity in the oceans.

155 - Shark teeth provide one of the richest sources of data about ancient marine life.

156 - The bite strategy of a shark combines tooth design with jaw mechanics for effective predation.

157 - Tooth wear and replacement rates can reflect a shark's feeding lifestyle and prey choice.

158 - Ancient and modern sharks together illustrate how tooth design has evolved under selective predation pressures.

159 - The continuous tooth renewal is a remarkable feature that sustains predation over a lifetime.

160 - Fossil teeth are often the most durable remnants of ancient sharks and help reconstruct past oceans.

161 - Shark teeth are among the most iconic features of predatory marine life.

3

SKIN LIKE SANDPAPER: SCALES, SPEED, AND SECRET ARMOR

Shark skin isn't just rough—it's a built-in engineering marvel. Covered in dermal denticles, tiny tooth-like scales that crest the body with ridges, their microstructure acts like a natural speed suit. This chapter dives into how these microscopic teeth on the skin help sharks glide faster, hide stealthily, and stay protected.

162 - Dermal denticles are tiny, tooth-like scales that cover shark skin.

163 - Each denticle has an outer enamel-like layer called enameloid and an inner dentin core.

164 - The denticles are anchored in the skin and grow with the shark.

165 - The crown of a denticle is connected to the base by a small stalk, similar to a tooth root.

166 - The ridges on denticles run from front to back, creating backward-pointing grooves.

167 - These ridges form microgrooves that channel water along the body.

168 - The microtexture reduces skin friction drag, helping swimming efficiency.

169 - The drag reduction saves energy, allowing sharks to travel longer distances.

170 - The rough texture of shark skin is due to the dense denticle ridges.

171 - Denticle patterns vary among species and across body regions.

172 - In fast-swimming species, denticles tend to be tall and strongly ridged.

173 - In other species, denticles are smaller or flatter, reflecting different lifestyles.

174 - The denticle system is ancient and present in early sharks.

175 - Fossil denticles help identify extinct shark species.

176 - Placoid scales are the category to which denticles belong.

177 - The enamel on denticles is hard and wear-resistant.

178 - The dentin core provides strength to each denticle.

179 - The mosaic arrangement of denticles creates the shark's unique skin texture.

180 - The armored skin with denticles protects against scrapes and minor injuries.

181 - The ridges on denticles can resist abrasion when rubbing along the seafloor.

182 - The skin around denticles contains mucus and protective layers.

183 - Fossil denticles preserve microstructure that helps trace shark evolution.

184 - The enamel and dentin composition in denticles is similar to tooth tissues.

4

SUPER SENSES: SMELL, HEARING, AND DETECTING A DROP OF DRAMA

Sharks rely on senses that go far beyond what most people imagine. This chapter dives into smell, hearing, and electroreception, revealing how these predators track prey, navigate murky waters, and strike with precision.

185 - Sharks smell with paired nostrils called nares on the snout that sample water without letting it flow into the mouth.

186 - Their olfactory system is extremely sensitive, capable of detecting trace chemical cues in the water.

187 - Olfactory lamellae inside the nostrils greatly enlarge the surface area for odor detection.

188 - Sharks can differentiate among odors and respond to scents associated with living prey.

189 - Odor plumes carried by currents can guide sharks toward potential meals.

190 - They can detect blood, amino acids, and other biologically important substances even when diluted.

191 - Different species show different levels of odor sensitivity, influenced by habitat and lifestyle.

192 - Electroreceptors help sharks detect the electrical activity of nearby prey.

193 - Electroreception can aid navigation by sensing the Earth's magnetic field for long-distance travel.

194 - These receptors can detect tiny electrical signals from prey beneath sand.

195 - Neuromasts in the lateral line detect vibrations and fluid flows caused by nearby animals.

196 - The lateral line helps sharks track prey even when water is cloudy or dark.

197 - The inner ear detects sound and assists with balance.

198 - Shark hearing is especially sensitive to low-frequency sounds typical of many prey and natural water movements.

199 - Low-frequency sounds from prey can draw sharks from a distance.

200 - Smell, electroreception, and the lateral line often work together to locate prey.

201 - In murky water, odor cues become more important for finding feeding opportunities.

202 - Sharks compare input from both nares to help pinpoint the odor source.

203 - Olfactory information travels to the brain and reaches the olfactory bulbs for processing.

204 - Some sharks have more olfactory tissue, increasing their scent-detection capacity.

205 - Sharks do not breathe through their nostrils; water flows over the gills for respiration.

206 - Odor cues help sharks identify productive feeding zones and avoid dead zones.

207 - Juvenile sharks rely on smell to locate small prey and safe nursery areas.

208 - The olfactory system is ancient and widespread among sharks and relatives.

209 - The cephalofoil may improve detection of chemical cues and electrical signals from a wider area.

210 - Odors associated with living prey excite hunting responses more than odors associated with decay.

211 - Rivers and estuaries can create chemical plumes that guide sharks toward feeding grounds.

212 - Some sharks migrate following chemical gradients in the water.

213 - Olfaction influences where sharks choose to hunt and rest.

214 - Smell remains active as sharks move, even at slow speeds.

215 - Odor cues can help sharks remember and revisit successful foraging spots.

216 - The brain integrates smell with vision, sound, and movement to plan a strike.

217 - Electroreception remains functional in darkness and murky water, making it crucial for hunting.

218 - The electroreceptors are arranged on the head and sometimes around the snout.

219 - Electrical signals from prey can be detected even when the prey is hidden or buried.

220 - Sharks detect living tissue through small electrical fields rather than just odor.

221 - The combination of chemical cues and electric fields provides a powerful hunting toolkit.

222 - The lateral line runs continuously along the body, giving real-time sense of water movement.

223 - Lateral-line sensing helps estimate the size, speed, and distance of a moving animal.

224 - Hearing complements other senses by picking up distant or faint noises.

225 - Sound travels faster and farther in water, providing timely cues to predators.

226 - The inner ear contains structures that translate sound into nerve signals.

227 - Hearing information helps sharks locate schools of fish by picking up collective movements.

228 - To hunt, sharks often use a multi-sensory search pattern that combines smell and vibration.

229 - Strong currents or odors that drift away from a prey source still provide directional clues.

230 - Saltwater quality can impact the spread of chemical cues and thus scent detection.

231 - Some sharks can maneuver in river systems, relying on smell and electroreception in fresh water.

232 - Different species rely on smell as part of a broader hunting strategy.

233 - The olfactory system is essential for locating prey in reefs, wrecks, and seagrass beds.

234 - Playful experiments test how sharks react to specific odors to map smell sensitivity.

235 - Scents released by injured prey are particularly attractive to many sharks.

236 - Olfactory processing is linked to appetite and feeding motivation.

237 - Odor cues can fade quickly, requiring sharks to follow fresh plumes for a successful hunt.

238 - The olfactory system influences habitat selection and migration patterns.

239 - Smell can be influenced by salinity and temperature, which change how chemicals disperse.

240 - Electrical cues help sharks detect buried prey as it sits on the seabed.

241 - Sharks' skin can sense water movement, providing another layer of sensory input.

242 - Scent trails and currents shape where sharks look first when searching for food.

243 - The olfactory system has multiple receptor types to detect various chemical classes.

244 - Hammerheads' sensory arrangement offers enhanced detection of smell across a broad area.

245 - Anatomical specialization highlights how different species optimize their senses.

246 - Scent-driven foraging reduces energy use by guiding sharks directly toward promising areas.

247 - Sensing environment helps sharks avoid dangerous areas by detecting predators' scents.

248 - Olfactory cues can be used to locate nurseries for young sharks.

249 - River bull sharks can smell their way through muddy estuaries.

250 - Odor detection plays a role in social interactions for some species.

251 - Odor detection is linked to learning and memory in sharks.

252 - The sense of smell is one of the oldest in vertebrate evolution.

253 - The navigation system in sharks includes smell, electricity, and magnetism.

254 - Sharks' sense of smell can help scientists identify crucial feeding zones for conservation.

255 - The electroreception sense is unique to cartilaginous fish among many aquatic vertebrates.

256 - Lateral line sensors are tuned to detect movement patterns of schools.

257 - Sharks can adjust their sensitivity to odors in response to environmental changes.

258 - Some sharks can smell certain compounds better at certain water temperatures.

259 - Odor gradients help create a mental map of the local ocean around a reef.

260 - The combination of senses results in rapid, efficient hunts.

261 - Scent cues help in predator avoidance by signaling the presence of larger predators.

262 - A shark's senses are used not only for hunting but also for habitat selection.

263 - The hammerhead's sensory network is among the most expansive of any shark.

264 - The sense of smell can help track scent trails in rivers and estuaries even when water is murky.

265 - Sharks can hunt during both day and night because of their multi-sensory toolkit.

266 - The sensory system supports a flexible hunting strategy across diverse habitats.

267 - Scientists study shark senses to design better conservation and human-wildlife guidelines.

268 - The senses help scientists interpret shark behavior in the wild.

269 - Scent detection can inform how sharks migrate along coastlines.

270 - The sense of sight is not needed for all hunts; smell and electroreception play major roles as well.

271 - The senses allow sharks to function as apex predators with tuned hunting abilities.

272 - The sensory system enables sharks to capture prey in open water and near the bottom.

273 - The olfactory system contributes to spatial awareness and orientation.

274 - River-dwelling sharks like bull sharks use smell to locate prey in murky waters.

275 - The senses help conservationists tag and track sharks by understanding their foraging cues.

276 - The sensory toolkit supports sharks' learning and adaptation to diverse habitats.

277 - There is still much to learn about how these senses work together in different species and environments.

5

ELECTRIC POWERS: THE SIXTH SENSE THAT FINDS HIDDEN PREY

Sharks sense more than sight and smell—they can feel the electric heartbeat of the ocean. In this chapter, we dive into the ampullae of Lorenzini, the jelly-filled channels that turn faint electrical fields into hunting clues. From bury-under-sand ambushes to lightning-quick strikes, electroreception reshapes the way sharks find hidden prey.

278 - The ampullae are gel-filled canals that connect to sensory cells in the shark's skin.

279 - These receptors detect the electrical fields produced by living animals.

280 - Electroreception is highly sensitive and can pick up very small electrical signals.

281 - The sense is most responsive to low-frequency electrical activity typical of muscle movement.

282 - Sharks compare signals across many ampullae to determine where a field is strongest.

283 - The cephalofoil helps hammerheads locate prey moving beneath the sand by sampling energy around a broad area.

284 - All sharks possess ampullae; the system is ancient in cartilaginous fish.

285 - The jelly inside the ampullae aids in transmitting electrical signals to the sensory nerves.

286 - Electroreception helps sharks hunt in turbid water where vision is poor.

287 - Ampullae density and distribution vary among species according to their hunting style and habitat.

288 - In dim light, electroreception can be a primary tool for locating prey.

289 - Signals from ampullae travel via cranial nerves to the brain for processing.

290 - Electroreception complements other senses like smell, sight, and the lateral line.

291 - The ocean contains background electrical noise that sharks must distinguish from prey signals.

292 - In labs, sharks respond to artificial electric fields presented around bait setups.

293 - The electric sense explains some fast, surprising shark attacks on seemingly calm targets.

294 - The snout region often contains the highest concentration of ampullae.

295 - The evolution of electroreception likely helped sharks exploit hidden or buried prey.

296 - The ampullae create a directional sense by comparing signal strength across receptors.

297 - Electroreception is especially useful for detecting prey beneath the sand or mud.

298 - The sense works even when visual cues are unavailable or unclear.

299 - Different shark species rely on electroreception to varying degrees depending on ecology.

300 - Electroreception is part of a shark's overall hunting toolkit, alongside vision and smell.

301 - Ampullae number and layout influence how well a species can sense weak fields at distance.

302 - The canals funnel signals to sensory cells that convert electrical activity into nerve impulses.

303 - The electric sense operates best at close to intermediate distances where fields are clearest.

304 - The brain integrates electroreceptive information with other sensory data to guide behavior.

305 - The electric sense can help sharks pinpoint prey even when water is murky or dark.

306 - Differences in signal strength across the head create directional information about the prey's location.

307 - Sharks can respond to a range of frequencies produced by prey depending on movement and species.

308 - Species with dense ampullae often hunt for prey that hides in substrate or sand.

309 - The name "ampullae of Lorenzini" honors a scientist who described them in the past.

310 - The jelly in the ampullae continually interacts with seawater to keep signals readable.

311 - Juvenile sharks also possess ampullae and begin using electroreception early in life.

312 - The sense supports rapid attacks by providing quick directional cues.

313 - Ampullae are primarily located on the snout, with additional receptors around the mouth and eyes in some species.

314 - Electroreception is a distinctive feature of sharks and rays among fish.

315 - Environmental factors like currents and turbidity can affect the clarity of electrical signals.

316 - The electroreception field has inspired engineers to design new underwater sensors.

317 - Polluted or changing water chemistry can alter the sensitivity of electroreceptors.

318 - The electroreception system is one of the oldest sensory systems in sharks.

319 - The electric sense shapes how sharks move and strike during predation.

320 - The ampullae help explain why sharks often ambush prey with sudden, precise turns.

321 - Electroreception is especially valuable for ambush predators that lie in wait.

322 - Ampullae form a sampling grid across the shark's head to detect electric fields from multiple directions.

323 - Regions around the eyes and nostrils may host additional ampullae in some species.

324 - The electroreceptive system remains a major focus of ongoing shark research.

325 - Other aquatic animals also use electroreception, but sharks have one of the most developed systems.

326 - The field patterns of living tissue differ from inanimate objects, helping sharks distinguish prey from rubble.

327 - Electric signals are detected even when the water is moving, though noise can complicate reading them.

328 - The ampullae connect to nerves that relay information to brain regions involved in sensory processing.

329 - Large sharks tend to have extensive electroreceptive networks to cover more area.

330 - Electroreception helps predict prey movement and plan a quick approach.

331 - The field around prey changes as animals swim and breathe, creating dynamic signals for sharks to chase.

332 - Scientists use computer models to study how electric fields spread around prey and predators.

333 - The electric sense can increase a predator's success rate in low-visibility habitats.

334 - The timing of electrical signals can influence when a shark decides to strike.

335 - Conservation researchers study electroreception to understand how sharks find food in changing oceans.

336 - The ampullae translate electrical signals into fast neural responses.

337 - Electroreception remains one of the most fascinating sensory modalities in nature.

338 - The sense is especially helpful for ambush hunters that lie camouflaged and wait for prey to come within range.

339 - Electroreception is a stable feature across most shark species, underscoring its importance.

340 - Some popular portrayals of sharks emphasize electroreception as a primary sense, though it works with others.

341 - A shark's own movement can create background electrical signals that it must account for.

342 - The electric signals from prey vary with size and activity, providing clues to sharks about what they are chasing.

343 - The combination of electroreception with other senses lets sharks adapt to a wide range of environments.

344 - The brain areas processing electroreception are still a focus of neurological study.

345 - The distribution of ampullae around the head creates a broad sampling field for detection.

346 - Electroreception is not visible to the naked eye but is a real, measurable sense.

347 - The evolutionary history of electroreception reveals long-standing adaptation to aquatic life.

348 - Researchers explore how electroreception influences predator–prey interactions in the ocean.

349 - The sense operates alongside the lateral line and other senses to guide behavior.

350 - Damage to the snout can temporarily reduce electroreception, illustrating the sensitivity of the system.

351 - The ability to sense electricity allows sharks to respond quickly to sudden movement in their environment.

352 - Electroreception has inspired bio-inspired designs for underwater robots.

353 - Not all electric signals are from prey; man-made electrical sources can create background noise or attract sharks depending on field.

354 - A diverse array of species relies on electroreception, from small ground sharks to large great whites.

355 - The ampullae are visible only when examining the head with careful tools; they are not obvious in the wild.

356 - Environmental changes in oceans can alter the effectiveness of electroreception by affecting water conductivity.

357 - The electroreceptive system provides a model for studying brain–sense integration.

358 - The presence of electroceptors has shaped the way scientists design field experiments.

359 - The "sixth sense" label reflects how powerful electroreception is as part of a shark's senses.

360 - Research into electroreception continues to reveal new details about sharks' hunting strategies.

361 - The ability to sense electricity extends to some rays and other cartilaginous fishes.

362 - Understanding electroreception helps scientists protect sharks by understanding their food needs and habitats.

363 - Electric sensing remains a striking example of sensory adaptation in the ocean.

364 - The electroreceptive system is an enduring source of wonder for ocean lovers and researchers alike.

6

GLOW-IN-THE-DARK, SEE-IN-THE-DARK: DEEP-SEA SHARK WEIRDNESS

Nighttime in the ocean becomes a stage for some of the planet's brightest hunters. This chapter dives into glow-in-the-dark and see-in-the-dark adaptations of sharks, from photophores and luminous patterns to giant eyes and slow, stealthy movements in the deep.

365 - Some deep-sea sharks glow in the dark using photophores—special light-producing organs.

366 - The light emitted by these sharks is usually blue or blue-green because those wavelengths travel well in seawater.

367 - A common use of glow is counter-illumination: ventral light helps conceal their silhouette when viewed from below.

368 - Lanternsharks are among the best-known glow-bearing sharks, with many species that glow in the deep.

369 - Photophores can appear as spots, bars, or lines along the body, and patterns can differ between species.

370 - In some species, photophores can be adjusted to glow more or less, enabling light-based signaling.

371 - Some photophores host luminous bacteria that produce light within the organ.

372 - Bioluminescence may help lure prey toward the shark's mouth in the darkness.

373 - Glow can also function as a signaling channel for potential mates or rivals.

374 - Ventral glow helps the shark stay hidden from predators or prey looking up from below.

375 - Deep-sea sharks often have relatively large eyes in relation to their body size to help detect the faint light.

376 - Some species have tubular eyes that help concentrate light in dim water.

377 - The retinas of many deep-sea sharks rely heavily on rod cells for sensitivity in low light.

378 - In darkness, other senses such as the lateral line and olfactory system become especially important.

379 - Bioluminescence is only one of several nocturnal adaptations that help these sharks hunt.

380 - Many glow-capable sharks migrate vertically, moving toward shallower depths at night to feed.

381 - The glow, combined with color patterns and body shape, can aid camouflage in certain environments.

382 - A large, oil-rich liver contributes to buoyancy and energy storage in the deep sea.

383 - The deep-sea environment is cold and high-pressure, shaping metabolism and glow efficiency.

384 - Bioluminescence has evolved multiple times across different shark lineages, illustrating convergent adaptation.

385 - Photophores can be located on the belly, sides, or tail, depending on the species.

386 - Light emission in photophores is produced by chemical reactions, sometimes powered by symbiotic bacteria.

387 - The brightness of photophores can vary with depth and physiological state, among other factors.

388 - Some glow displays are steady, while others pulse or flash as signals.

389 - The glow is most apparent in complete darkness when ambient light is minimal.

390 - Photophore patterns can help scientists identify and classify deep-sea sharks in the field.

391 - The presence of glow indicates a predator that relies, at least in part, on light-based strategies in dim environments.

392 - Photophores are typically embedded in the skin rather than being external appendages.

393 - The production of light requires energy, so glow is used strategically rather than constantly.

394 - Bioluminescent sharks can evoke an eerie, otherworldly appearance when seen in the dark.

395 - The glow can contribute to a shark's distinctive silhouette, aiding researchers in recognition.

396 - Some photophores glow in patterns that form recognizable silhouettes along the body.

397 - Photophore brightness can vary with recent meals and overall health.

398 - The deep-sea environment supports an interconnected ecosystem of light-producing organisms.

399 - Luminescence throughout the deep sea is used for predation and defense in complex ways.

400 - Light production is one of many tools a shark may use to survive in the deep.

401 - The presence of luminescence in a region can influence how other animals behave nearby.

402 - Photophore placement and brightness often reflect a shark's ecological niche.

403 - Studying glow in sharks involves submersibles, remotely operated vehicles, and specialized cameras.

404 - Glow can alter how nearby prey and predators respond at close range.

405 - Researchers sometimes describe deep-sea glow as a natural underwater display of bioluminescence.

406 - The luminous system is an excellent example of deep-sea adaptation and diversity.

407 - Bioluminescence is more common at depths where residual light is scarce but present.

408 - In some species, photophores extend across broad areas of the body.

409 - The glow can assist in social interactions among individuals in low light.

410 - Light production is a visible cue that helps scientists study shark behavior in darkness.

411 - The presence of glow suggests that light-based strategies can be advantageous in dim environments.

412 - Photophore can contribute to a shark's personal light signature used in field studies.

413 - The glow may be used during mating displays under the cover of night.

414 - The brightness of photophores can be manipulated to avoid attracting non-target organisms.

415 - Bioluminescence in sharks is part of the broader deep-sea lightscape shared with other organisms.

416 - Some luminous sharks occupy midwater zones where both dark and light mix.

417 - The glow provides a visual cue that helps researchers track movement in the dark.

418 - The luminous organs are typically small but highly efficient.

419 - Bioluminescence interacts with water conditions, affecting how far light travels.

420 - The glow is easier to study with underwater imaging techniques that avoid flashing lights.

421 - The deep-sea glow demonstrates how life adapts to living in perpetual shade.

422 - Photophores can be used to deter rivals by signaling presence or strength.

423 - The glow can influence prey behavior, potentially guiding them toward a predator's mouth.

424 - Bioluminescence embodies the incredible energy budgets of deep-sea predators.

425 - The luminous system has captured popular imagination, becoming a symbol of deep-sea mystery.

426 - Photophores can differ in brightness across the body, creating dynamic overall glow.

427 - The glow may be seasonally variable in some species, tied to reproduction or feeding cycles.

428 - The deep-sea glow is one reason some sharks look "alien" to us.

429 - Some researchers study luminous sharks to understand how light travels and diffuses in the ocean.

430 - The glow can be an advantage in foraging by making prey more visible to the hunter.

431 - Photophores can provide an outward sign of a shark's health and vitality in some contexts.

432 - The deep-sea light world interacts with ocean currents, influencing light distribution.

433 - Luminescent sharks illustrate how evolution can tailor biological light for survival.

434 - Photophore visibility to divers depends on camera settings and light levels.

435 - The glow inspires awe and curiosity in audiences who encounter it in ocean documentaries.

436 - Some luminous sharks are small and use their glow to complement stealth hunting.

437 - The energy costs of light production influence a shark's daily behavior.

438 - The deep sea is not uniformly dark; glows add to its complexity.

439 - Luminescence can shape predator-prey dynamics by changing how prey respond to a glowy hunter.

440 - In some settings, light helps researchers map the distribution of luminous sharks.

441 - The glow demonstrates the connection between biology and physics in nature.

442 - Bioluminescence reveals how life can thrive in one of Earth's most extreme habitats.

443 - The luminous system gives scientists clues about the evolutionary history of sharks.

444 - Photophores can sometimes be visible to divers with specialized lighting.

445 - The glow exemplifies how marine life uses light to navigate unseen worlds.

446 - Some luminescent sharks have photophores that appear on both sides of the body.

447 - The glow adds to the drama of deep-sea ecosystems in educational media.

448 - The deep-sea glow may be used in combination with other camouflage strategies.

449 - Light production is a defined trait among certain deep-sea sharks, not universal.

450 - The glow can help divers observe shark behavior in night-time field studies from boats.

451 - The luminous world of sharks is part of the natural history of marine science.

452 - Photophores provide a visible signature for scientists to track populations.

453 - The glow's physics illustrate how light interacts with seawater, scattering and absorbing at different rates.

454 - Bioluminescence in sharks adds to the diversity of predator strategies in the deep.

455 - The glow helps illustrate how life adapts to darkness and pressure.

456 - Photophores can be a key feature in taxonomic descriptions of luminous sharks.

457 - The glow highlights the unseen life that thrives where daylight never reaches.

458 - Scientists continue to uncover new luminous species in expeditions around the world.

459 - The glow reveals how deeply interconnected deep-sea ecosystems are through light.

460 - The luminous sharks remind us that the ocean can be full of color even in the dark.

461 - Evolution has crafted a world where light itself becomes a weapon, a lure, and a language.

462 - Glow-in-the-dark sharks stand as vivid ambassadors for the strange beauty of the deep sea.

7

SHARKS THAT WALK, WIGGLE, AND PULL OFF WEIRD MOVES

Sharks move in some of the ocean's most surprising ways, from tide-pool walkers to tail-whip speedsters. In this chapter, you'll meet bottom crawlers, glow-in-the-dark ambushers, and lightning-quick hunters, all showing how movement fuels their amazing predatory lives.

463 - Epaulette sharks can walk on their fins along the ocean floor.

464 - They can move between connected tide pools when the water recedes, effectively crawling across the reef.

465 - Walking sharks belong to the Hemiscyllium genus, a small group of reef-dwelling species.

466 - Their walking gait is slow and deliberate, not the fast swimming you see in sleek pelagic sharks.

467 - They lift their bodies slightly and shift weight from one fin to the other to crawl along the bottom.

468 - Walking helps them hunt in shallow, rocky areas where swimming would churn up sand.

469 - They often hide among rocks and corals, stepping stealthily to surprise prey.

470 - Walking sharks feed on small fish, crustaceans, and other invertebrates found in tide pools and shallow reefs.

471 - Their skin is rough and sandpaper-like, helping them grip the seafloor as they move.

472 - They are most commonly found in tropical Indo-Pacific waters.

473 - Wobbegongs are carpet sharks that camouflage themselves on reefs and in kelp beds.

474 - They move slowly along the bottom using their broad pectoral fins to glide rather than chase.

475 - When prey comes within reach, they lunge forward with a sudden, precise strike.

476 - Wobbegongs can stay nearly motionless for long periods, conserving energy.

477 - Their fringe-like skin flaps help them blend into the reef and confuse predators.

478 - Thresher sharks are famous for their extra-long tail fin.

479 - They use their tail as a whip to herd schooling fish into striking distance.

480 - A tail swipe can sweep several fish into a tight group for an easy feeding opportunity.

481 - Threshers swim with a smooth, powerful, tail-driven motion.

482 - Hammerhead sharks have a wide cephalofoil head that looks like a hammer-shaped shield.

483 - That wide head improves their ability to turn quickly and stalk prey.

484 - Hammerheads often form schools, especially when they are young.

485 - When hunting, hammerheads sweep from side to side, pinching prey against coral or the ocean floor.

486 - The head shape helps reduce blind spots so they can see prey from more angles.

487 - They use powerful tail strokes and a streamlined body to surge through the water.

488 - Shortfin makos can leap clean out of the water during hunts or when excited by a boat.

489 - Great white sharks perform rapid bursts and can breach the surface when chasing prey like seals.

490 - Their torpedo-shaped bodies reduce drag and help them accelerate quickly.

491 - They venture into rivers to hunt fish, dolphins, and other prey in murky water.

492 - Bull sharks' muscular bodies help them strike fast in tight spaces.

493 - They breathe by gently pumping water over their gills while they rest.

494 - Nurse sharks can remain stationary while water flows over their gills to ventilate.

495 - Frilled sharks move with a sinuous, eel-like motion suited to their deep-water habitat.

496 - They dwell in deep, dark oceans and feed on smaller fish and squid.

497 - They swim with a slow, sinuous motion that suits their deep-sea habitat.

498 - Bioluminescence helps lanternsharks camouflage by matching faint downwelling light.

499 - Some lanternshark species have photophores in patterns that may help communicate with mates.

500 - The glowing patterns can attract prey or confuse predators at depth.

501 - Bioluminescence is most common among small, deep-sea sharks.

502 - The pectoral fins provide lift, helping keep the body level in the water during movement.

503 - The tail fin (caudal fin) is the main engine for fast swimming in many sharks.

504 - Some bottom-dwelling sharks use short, creeping movements to reposition themselves without creating big waves.

505 - The swimming style of a shark depends on its habitat and prey.

506 - Some sharks use a carangiform pattern, with most propulsion from the tail and little body movement.

507 - Other sharks show a slow, eel-like undulation when feeding in tight spaces.

508 - The widened sensory field lets them detect prey in cluttered environments.

509 - Many deep-sea sharks move slowly to conserve energy in limited food environments.

510 - Some sharks can glide by spreading their pectoral fins to ride cross-currents.

511 - When moving in schools, some species maintain tight, synchronized spacing.

512 - A quick dart forward can start a feeding frenzy by scattering prey.

513 - The ability to turn sharply helps predators cut off escape routes for prey.

514 - This sense lets them detect muscle movement in prey even when it is out of sight.

515 - Some sharks coordinate movement with currents to conserve energy.

516 - They may ride currents with gliding motions before diving to hunt.

517 - Hovering behavior is achieved by balancing lift with their fins when not actively swimming.

518 - Some species can stay at a fixed depth and drift slowly, conserving energy.

519 - During feeding frenzies, sharks can remain close together while grabbing prey in quick succession.

520 - Some sharks rely on ambush rather than chasing to catch prey, using movement only when necessary.

521 - The speed of a hunting strike is often the difference between success and failure.

522 - The differences in swimming styles reflect adaptation to prey types and habitats.

523 - Once a prey item is detected, some sharks launch a fast, short, high-velocity strike.

524 - The body shape of a shark influences its maximum acceleration and turning radius.

525 - Some sharks can shift from cruising to sprinting in a fraction of a second.

526 - In murky river water, movement can become more reliant on sensing rather than sight.

527 - The prey's reaction after a bite can influence the predator's next move.

528 - The energy cost of movement is a key factor in how sharks hunt.

529 - Juvenile sharks often practice movement skills in shallow, safe waters.

530 - Some sharks change their gait as they grow, moving differently as adults.

531 - Variety in movement styles shows how sharks adapted to habitats from rivers to deep sea.

532 - The walking sharks are iconic for their unusual seafloor locomotion.

533 - The ability to walk enables walking sharks to exploit microhabitats unreachable to pelagic sharks.

534 - Bioluminescent sharks rely on light to survive and navigate in the dark depths.

535 - The glow helps them hide against predators while they move.

536 - Some sharks have broad, flat heads that help them move along the seabed with stability.

537 - A few sharks can propel themselves briefly upward to escape threats.

538 - Movement in sharks is shaped by a mix of anatomy, physiology, and behavior.

539 - The combination of fin shape and body length determines how swiftly a species can move.

540 - Some sharks perform rapid lateral movements to outmaneuver prey.

541 - The walking shark's stride is a memorable example of unusual shark movement.

542 - The slow, creeping motion of bottom-dwellers can look almost like stalking.

543 - The flick of a tail can scatter small fish, triggering a feeding frenzy.

544 - A hunter like the mako or great white can catch prey in a single, rapid rush.

545 - Some sharks hover or glide briefly before striking to gauge the best angle of attack.

546 - The way sharks move is closely tied to what, where, and when they hunt.

547 - Movement patterns can shift with water temperature and prey availability.

548 - The senses and movement work together to create an efficient predator.

549 - Walking sharks live in coral reefs across the western Pacific and Indian Oceans.

550 - Sharks constantly surprise us with new movements as they adapt to changing oceans.

8

TINY TERRORS TO OCEAN GIANTS: RECORD-SETTING SIZES

From pocket-sized swimmers to towering ocean giants, size is the headline here. This chapter dives into record-setting lengths, growth stories, and the real science behind the biggest—and smallest—sharks.

551 - The dwarf lanternshark is the smallest known shark, reaching about 17 centimeters in length.

552 - In the world of sharks, several species stay well under 20 centimeters, showing how tiny some predators can be.

553 - The cookiecutter shark grows to roughly 50 centimeters, giving it a compact but fearsome bite.

554 - Bioluminescent sharks glow in the dark to blend with dim ocean light and confuse prey, a handy trick for small and large alike.

555 - Some deep-sea sharks use their glow patterns to communicate with mates and rivals.

556 - Whale sharks are the largest living fish, commonly reaching around 12 meters in length.

557 - Basking sharks are the second-largest living sharks and can reach around 12 meters.

558 - Great white sharks can reach lengths of about 6 meters in the wild.

559 - Tiger sharks can grow to around 5 meters, with some larger individuals.

560 - Hammerhead sharks, especially the great hammerhead, can reach around 6 meters.

561 - Megamouth sharks typically reach about 4 to 5 meters in length.

562 - Goblin sharks can reach roughly 3 to 4 meters.

563 - Frilled sharks reach about 2 meters in length.

564 - Greenland sharks can reach lengths around 6 meters, though few reach this size.

565 - Megalodon is estimated to have reached about 15 to 18 meters in length.

566 - Fossils show giant Otodus species grew to lengths comparable to or larger than megalodon.

567 - The largest extinct sharks exceeded many living species in size, illustrating a long history of giant predators.

568 - Newborn sharks vary in size by species; some are only a few centimeters long at birth.

569 - Growth rates vary widely: some species gain centimeters quickly in youth, then slow.

570 - Whale sharks grow slowly and can take decades to reach full size.

571 - Juvenile sharks often spend years in protected nursery areas before joining adults.

572 - The biggest sharks require large, productive habitats to access enough prey.

573 - Size influences prey choice, with bigger species often tackling larger prey items.

574 - The long tails of some large sharks, like threshers, contribute to their impressive speed.

575 - The response to prey depends on size, with some large sharks making rapid strikes.

576 - The tails of large sharks can be longer than their bodies in some cases, aiding propulsion.

577 - The body shape of a giant filter feeder like a whale shark is designed for endurance, not sprinting speed.

578 - The largest sharks are often found in coastal zones where prey is plentiful, before moving into open water.

579 - The largest sharks typically reach sexual maturity only after many years.

580 - A large female's offspring survival can depend on the size and health of the mother.

581 - The anatomy of giant sharks, including their cartilaginous skeletons, supports buoyancy and growth.

582 - The sense of electroreception helps giants locate prey at depth without relying on sight.

583 - The growth trajectory of a giant shark reflects the availability of prey over time.

584 - Some giant sharks migrate long distances in search of food-rich waters.

585 - The digestive system of huge sharks is adapted to process vast amounts of water while filtering.

586 - Fossil records show ancient oceans hosted giants that dwarfed many living species.

587 - The biggest sharks are not necessarily the strongest at every moment; size is just one advantage.

588 - The presence of enormous sharks can shape the entire ecosystem by controlling prey populations.

589 - The great white's size helps it ambush prey near the surface.

590 - The length of a shark does not always predict its bite force; bite mechanics matter.

591 - The teeth of large predators are continuously replaced, enabling sustained feeding capability.

592 - The jaws of big sharks are anchored and powerful to secure large prey.

593 - The distribution of giant sharks has shifted with climate and prey populations.

594 - Giant sharks are more common in warmer, nutrient-rich waters that support abundant prey.

595 - Giant sharks can reach their maximum size only after many years of growth.

596 - Fossil megalodon teeth found around the world indicate broad distribution and large size.

597 - The body plan of giant sharks is the result of millions of years of evolution.

598 - The filter-feeding giants rely on plankton and small organisms to reach massive sizes through sheer numbers.

599 - The large organs in the abdomen help with buoyancy in giant sharks.

600 - Some giants migrate long distances to take advantage of seasonal prey swings in the ocean.

601 - Largest sharks can inhabit both shallow and deep waters, depending on species and season.

602 - Record sizes are often clarified by careful measurement and verification from multiple data sources.

603 - The largest living predator by mass is the whale shark, thanks to its enormous girth.

604 - Regional differences can lead to some populations of large sharks being larger on average than others.

605 - The average length of a great white varies by region, with some populations growing larger than others.

606 - The length-to-weight ratio changes as sharks mature and accumulate fat and oils in their livers.

607 - To stay sharp, sharks continually shed and replace teeth, keeping their jaws ready for big meals.

608 - Some large sharks have broad snouts that help them scoop prey from the water column.

609 - Energy demands of a large body require abundant food, influencing movement and feeding patterns.

610 - Big sharks often use ocean currents to travel long distances between feeding and breeding grounds.

611 - Extinct giant sharks left behind fossil evidence that still stuns scientists today.

612 - Megalodon teeth found in many places indicate it ranged widely during its time on Earth.

613 - Some giant sharks evolved special tooth shapes for gripping large prey like whales.

614 - The discovery of giant shark fossils continues to expand our understanding of prehistoric oceans.

615 - In modern oceans, the largest sharks are often filter feeders rather than active hunters of big prey.

616 - The mass of a whale shark can rival the weight of several cars, depending on its size.

617 - The basking shark's gaping mouth and thousands of baleen-like plates help it skim enormous amounts of water for food.

618 - Even the biggest sharks rely on steady prey to survive long-term.

619 - A shark's growth is influenced by temperature: warmer waters can speed up metabolism and growth.

620 - Female giants are often larger than males, a common pattern among big sharks.

621 - A juvenile's size at birth can influence lifetime growth strategy and longevity.

622 - Some large sharks travel across entire oceans, spanning thousands of kilometers as they mature.

623 - Size can influence how prey is processed and how prey is captured, depending on the species.

624 - The tail of a large shark can contribute significantly to propulsion during bursts.

625 - The mouth width of a large filter feeder like a basking or whale shark is immense, aiding feeding efficiency.

626 - A shark's size can influence its social structure, with some large species showing more solitary behavior.

627 - The record sizes of sharks are often revised with new measurements and discoveries in paleontology.

628 - The largest living sharks by mass are typically giants like the whale shark and basking shark, followed by top-level predators like the great white and megamouth.

629 - Some extinct sharks achieved sizes that modern researchers still find astonishing.

630 - Scientists use fossils, captured specimens, and imaging to estimate the size and bite capabilities of giants.

631 - The study of shark size helps scientists model ancient oceans and modern food webs.

632 - The ocean's climate and currents influence where large sharks can grow to record sizes.

633 - Record-size sharks inspire conservation efforts by highlighting the ecological importance of large predators.

634 - Technological advances let researchers estimate size and movement of sharks in the wild without heavy handling.

635 - The biggest sharks are not only about raw size— they bring specialized feeding modes that maximize energy intake.

636 - The evolution of gigantism in sharks reflects deep ocean productivity and prey diversification over millions of years.

637 - The study of size extremes in sharks connects biology, geology, and climate science in fascinating ways.

638 - The dramatic differences between tiny and giant sharks showcase the extraordinary range of life strategies in the sea.

639 - Size extremes in sharks remind us that the ocean still holds many mysteries waiting to be uncovered.

640 - From micro-sized to mega-sized, shark body plans reveal a continuum of design optimized for survival across vast oceans.

641 - Record-sized giants continue to captivate scientists, educators, and curious readers alike.

642 - The curiosity about shark size helps foster interest in marine conservation and STEM learning.

643 - By studying size, we gain insights into how evolution crafts incredible diversity among predators.

644 - Tiny sharks and huge sharks together illustrate the ocean's astonishing range of life.

645 - The giant predators' sheer size influences their role as apex defenders of marine ecosystems.

646 - Fossil giants and living giants both teach us about ancient climate, prey availability, and evolutionary innovations.

647 - Understanding size helps explain why sharks occupy so many ecological niches across the world.

648 - The extreme ends of shark size show how different life histories can be successful in the same ocean.

9

STRANGE BODIES DEPARTMENT: HEADS, FINS, AND OTHER BIZARRE PARTS

From hammerheads to goblin jaws, this chapter dives into the strange bodies that make sharks some of the ocean's most unusual engineers of evolution. Explore bizarre heads, fins, and other odd parts, and learn how these designs help sharks hunt, hide, and thrive in a world of giants and deep-dark waters.

649 - The hammerhead's head placement gives the eyes a broad, nearly panoramic view around the head.

650 - The cephalofoil helps hammerheads maneuver around corals and along the seabed with greater turning precision.

651 - Sawsharks have a long rostrum with teeth on both sides that resembles a saw blade.

652 - They slash through schools of fish with the rostrum to stun prey and create openings for the jaws.

653 - The teeth along the rostrum aid in trapping prey and guiding it toward the mouth.

654 - They rely on stealth and ambush rather than speed to catch prey.

655 - Wobbegongs have barbels near the mouth that help sense vibrations in murky water.

656 - The jaw extension allows rapid capture of prey in tight, deep-sea spaces.

657 - The goblin shark's snout houses electroreceptors to detect prey's electrical signals on the sea floor.

658 - Their pale skin helps them blend into the dim, deep sea environment.

659 - Frilled sharks have eel-like bodies and a frilly crown of gill slits around the neck.

660 - They belong to one of the oldest shark lineages, with relatives appearing in fossils hundreds of millions of years ago.

661 - Frilled sharks can coil into crevices to ambush prey from hidden angles.

662 - They inhabit deep, cold waters and can grow very large, though they are rarely seen by people.

663 - Sixgill sharks have elongated bodies built for slow, steady cruising in the dark depths.

664 - They whip the tail through schools to stun prey and disrupt schooling behavior.

665 - The tail whip can generate powerful shocks that disorient prey in an instant.

666 - Bonnethead sharks are a hammerhead relative with a shorter cephalofoil and a distinctive diet.

667 - Bonnetheads have been observed nibbling seagrass, showing plant matter in some sharks' diets.

668 - Seagrass in bonnethead diets demonstrates dietary flexibility in sharks.

669 - The cephalofoil distributes sensory input widely, increasing detection of prey signals.

670 - Hammerheads often migrate in groups, using their head shapes to help with maneuvering and spacing.

671 - The hammerhead family is scientifically classified as Sphyrnidae.

672 - Goblin sharks inhabit the deep sea where light is scarce, contributing to their pale, translucent appearance.

673 - Their jaws can extend rapidly to grab prey in tight spaces.

674 - Their teeth are small and needle-like, suited for grabbing slippery prey rather than tearing.

675 - They have large mouths capable of swallowing prey in a single gulp.

676 - Cookiecutter sharks emit light from photophores to attract larger animals toward their bites.

677 - They leave circular bite wounds on larger hosts, a signature feeding pattern.

678 - Photophores along a cookiecutter's body enable counter-illumination and lure.

679 - Cookiecutter sharks are small but effective deep-sea predators with a dramatic bite style.

680 - They often swallow prey head-first to reduce injury from teeth during feeding.

681 - Whale sharks are filter feeders with enormous mouths and broad, flat heads.

682 - They filter plankton as water passes through their gill rakers.

683 - They glide with mouths open to maximize water intake for filtration.

684 - Bioluminescence in lantern sharks is typically blue-green and concentrated along the underside.

685 - Deep-sea sharks like sixgill and goblin rely on slow metabolism to survive in nutrient-poor environments.

686 - Greenland sharks have very long lifespans and inhabit cold Arctic waters.

687 - They mature late and grow slowly, contributing to a protracted life history.

688 - Greenland shark flesh can be toxic if not properly processed, limiting human consumption.

689 - Dorsal fins provide stability and balance during swimming and turning.

690 - The size and shape of dorsal fins vary by species and habitat, reflecting different needs for stability and display.

691 - Pectoral fins act as lifting surfaces and help with precise steering and buoyancy control.

692 - The tail fin (caudal fin) provides most of a shark's thrust and varies in shape between species.

693 - Many sharks possess heterocercal tails that create lift as they swim.

694 - The sense of smell in sharks is extraordinarily acute, enabling detection of tiny chemical cues from far away.

695 - Fin shapes and sizes reflect adaptations to different hunting strategies and environments.

696 - Head shapes in sharks have evolved to optimize senses and feeding methods across habitats.

697 - Jaw mechanics allow fast, powerful bites and efficient prey processing.

698 - Rostral sensory organs help identify prey in complex underwater environments.

699 - The mouth position and feeding style vary across species ranging from subterminal to terminal mouths.

700 - The tail is a primary propulsion engine for many fast-swimming sharks and can vary in its thrust and efficiency.

701 - Effective predation relies on timing, stealth, and strike accuracy in many shark species.

702 - Hammerhead morphology has been a strong signature of a predatory niche in shallow and reef-associated habitats.

703 - Frilled sharks illustrate how ancient predation strategies persist in modern oceans.

704 - Sixgill sharks exemplify deep-water adaptations that enable life in near-freezing, low-oxygen zones.

705 - Camouflage and patterning help many sharks remain unseen while waiting for prey.

706 - Bioluminescent organs in deep-sea sharks demonstrate convergent evolution for life in darkness.

707 - Deep-sea life drives remarkable variations in heads, fins, and sensory systems among sharks.

708 - The head-and-fin diversity in sharks highlights how natural selection crafts specialized tools for different niches.

709 - The strange designs of sharks show that form follows function even in the ocean's apex predators.

710 - The deep-sea realm hosts a hidden zoo of head shapes, jaw mechanisms, and luminous features among sharks.

711 - The evolution of fins has produced a spectrum from broad, stable fins to slender, fast-moving fins optimized for speed and maneuverability.

712 - The hammerhead's silhouette is instantly recognizable, signaling a unique predatory approach across multiple species.

713 - Frilled sharks' slow and stealthy approach demonstrates that speed is not always required for successful predation.

714 - Deep-sea sharks reveal how energy efficiency shapes anatomy in ecosystems with limited food resources.

715 - The interplay of senses, jaws, and fins creates a toolkit that lets sharks exploit a wide range of prey and habitats.

716 - The strange designs of sharks remind us that evolu-

tion can turn almost any body part into a specialized weapon.

717 - Deep-sea occlusions and reef edges alike host sharks with surprising adaptations that challenge our expectations of what a predator should look like.

718 - The world of sharks proves that even the most fearsome hunters can be masters of camouflage, patience, and precision.

719 - The diversity of head and fin forms across sharks shows that there is a niche for every ocean habitat, from coral reefs to the deepest trenches.

720 - The study of shark anatomy continues to reveal clever solutions to hunting, hiding, and surviving in environments that test the limits of life.

721 - This chapter celebrates the quirky, the creepy, and the incredible parts that make sharks a cornerstone of marine biology.

THE MENU: WHAT SHARKS EAT
(IT'S NOT JUST "ANYTHING")

Sharks don't have a single menu. In this chapter you'll discover the surprising ways different sharks eat, from plankton-filtering giants to deep-sea ambush specialists.

722 - Whale sharks and basking sharks are filter feeders that take in water and trap plankton with their gill arches.

723 - Megamouth sharks are filter feeders that use their enormous mouths to capture plankton in deep water.

724 - Great white sharks primarily hunt seals and sea lions, but they also eat fish and squid.

725 - Tiger sharks sometimes ingest non-food items like debris found at sea, though this is rare.

726 - Bull sharks can live in freshwater and rivers, hunting fish, dolphins, and other sharks in these habitats.

727 - Bull sharks hunt in shallow, murky waters where prey are abundant.

728 - Hammerhead sharks use the wide cephalofoil to sense prey and to pin down flat-bodied animals like stingrays.

729 - Great hammerheads commonly hunt stingrays on the seabed.

730 - Thresher sharks often prey on schooling fish such as sardines and anchovies.

731 - Shortfin mako sharks are among the fastest predators and chase tuna, swordfish, and other large fish.

732 - Blue sharks feed on pelagic fish and squid and can travel far across the open ocean.

733 - Lemon sharks forage around shallow bays and mangroves, feeding on stingrays and small fish.

734 - Leopard sharks hunt small fish and crabs along sandy bottoms in temperate zones.

735 - Blacktip sharks hunt along shorelines, feeding on small fish, squid, and crustaceans.

736 - White-tip reef sharks prey on small reef fishes and octopus within coral ecosystems.

737 - Silky sharks chase pelagic fish and squid in offshore waters.

738 - Oceanic whitetip sharks are opportunistic feeders that eat fish, seabirds, and squid, and they scavenge when available.

739 - Greenland sharks are slow-moving and often scavengers, capable of eating large carcasses.

740 - Frilled sharks are deep-sea ambush predators that feed on squid and fish with needle-like teeth.

741 - Goblin sharks dwell in the deep sea and feed on small fish and squid with extendable jaws.

742 - Angel sharks lie on the sea floor and ambush prey such as flatfish, rays, and crustaceans.

743 - Sawsharks use their long, serrated snouts to slash prey and capture small fish and crustaceans.

744 - Port Jackson sharks have beak-like teeth that crush mollusks and crustaceans.

745 - The teeth of sharks come in many shapes, with slender teeth for gripping fish, flattening teeth for crushing shells, and serrated teeth for tearing flesh.

746 - Bottom-during sharks rely on suction feeding to pull prey from the sand.

747 - Reef sharks hunt small reef fishes, octopus, and crustaceans around coral habitats.

748 - Some sharks hunt at the edges of reefs, where prey are abundant.

749 - Some sharks chase schools of fish to break up the school and feed.

750 - Thresher sharks' tails are extremely long relative to their bodies.

751 - The diet of a shark can shift with age, from invertebrates to larger prey.

752 - Some sharks feed primarily on squids, especially in deeper waters.

753 - Some sharks feed on sea turtles and their hard shells by biting near the carapace.

754 - Some sharks feed on crabs and lobsters by cracking their hard shells with specialized teeth.

755 - Reef sharks often feed around dawn and dusk when prey is more active.

756 - Pelagic sharks feed on schools of fish and squid as they migrate through open ocean.

757 - Riverine sharks feed on fish and invertebrates inhabiting freshwater environs.

758 - Deep-sea sharks rely on squid and other deep-water prey due to the scarcity of prey in the deep sea.

759 - The prey selection of sharks is influenced by prey density, season, and habitat.

760 - Some large sharks scavenge on the remains of dead animals at the surface.

761 - The energy value of prey influences how much time a shark will invest in hunting.

762 - Young sharks may rely on invertebrates before taking larger prey later in life.

763 - Teeth are continually replaced as they wear down from feeding.

764 - Shark senses can guide them to prey even when visibility is poor.

765 - A shark's bite can quickly disable prey, allowing easier consumption.

766 - Some sharks exhibit nocturnal feeding, taking advantage of prey that move more actively at night.

767 - Cookiecutter wounds leave distinctive circular scars on larger animals, aiding species identification.

768 - Crushing teeth in bottom-dwelling sharks help crack shells of crabs and mollusks.

769 - Some reef sharks feed on octopus and other soft-bodied prey.

770 - Certain sharks supplement their diet with jellyfish when other prey is scarce.

771 - Diet can vary widely even within a species depending on location and season.

772 - Juvenile sharks tend to feed on smaller prey to grow safely.

773 - Adult predators often expand their diet to larger prey items as they mature.

774 - Some sharks scavenge on shipwrecks or dead whales when encountered.

775 - The size and energy needs of a shark influence how aggressively it hunts.

776 - Being able to swallow prey whole benefits smaller or younger sharks with limited jaws.

777 - Some sharks navigate by following prey movement across the water column.

778 - Floating objects and upwellings attract prey schools that predators use to feed efficiently.

779 - Sharks can display selective feeding, preferring prey that maximizes energy gained per effort.

780 - The diet of a shark helps shape its role in the ecosystem, from herbivorous-like grazing in some niches to apex predation in others.

781 - Feeding behavior contributes to the dynamic balance of reef and open-ocean communities.

782 - The feeding habits of sharks reveal their incredible diversity, from filter feeders to top predators.

783 - Sharks' diets and hunting strategies are tightly linked to their body shapes and habitats.

784 - Overall, sharks occupy a wide range of dietary niches beyond the cliché of chasing anything that moves.

11

FEEDING FRENZY FACTS: FROM CALM CRUISING TO CHAOS

Feeding frenzies turn sharks from patient cruisers into rapid, spectacular predators, but real feeding is more studied than sensational. This chapter dives into how group feeding starts, how sharks avoid harming each other, what triggers a frenzy, and how movie myths compare with real behavior. Get ready for a dive into the wild dynamics of a feeding event.

785 - A feeding frenzy typically starts when prey is wounded or distressed, creating a blood trail that draws multiple sharks to the scene.

786 - Sharks detect blood and other chemical cues in very low concentrations, helping them locate a feeding patch from kilometers away.

787 - Visual cues from other sharks feeding can attract additional predators to the same prey patch.

788 - When several sharks encounter the same patch, competition escalates quickly, turning a simple kill into a frenzy.

789 - The first shark to bite often anchors the feeding by grabbing a large chunk and deterring newcomers.

790 - A bite-and-release pattern helps sharks conserve energy during a fast, high-energy feeding event.

791 - Most bites are short and precise, not the long, tearing chomps seen in movies.

792 - Some sharks swallow large bites whole and then reorient for another strike.

793 - Fast, powerful strikes are common in predatory sharks and can happen in a fraction of a second.

794 - Suction and jaw pressure work together to pull prey into the mouth during a bite.

795 - Teeth are shaped to cut and pierce prey, enabling rapid dismemberment when prey struggles.

796 - A prey patch can sustain several sharks for minutes or even hours, depending on prey size and density.

797 - The velocity of water and prey motion helps generate turbulence that draws in nearby sharks.

798 - Heavy blood loss from wounded prey can prolong a feeding session and attract scavengers.

799 - Seasonal migrations and prey fish runs can spark seasonal feeding frenzies along coastlines.

800 - Prey schooling behavior can complicate feeding by presenting many targets in a small area.

801 - The presence of other predators, such as dolphins or larger sharks, can alter how a feeding event unfolds.

802 - A dominant individual may control access to a large bite, influencing how others participate.

803 - Not every shark at a feeding site bites, and participation varies with hunger, size, and opportunity.

804 - Friction with submerged objects or gear can disrupt a feeding event and cause some sharks to pause.

805 - The energy payoff of a successful bite drives sharks to maintain a rapid feeding tempo.

806 - Some species use their broad snouts or body position to pin prey and reduce escape attempts.

807 - Mako and great white sharks can deliver blisteringly fast bites that seize prey almost instantaneous ly.

808 - Hammerhead species may use their head shape to stun or restrain prey during a frenzy.

809 - Sharks often bite and release rather than hold on, allowing more individuals a chance to feed.

810 - Injured prey are often torn into smaller pieces that can be swallowed more easily.

811 - Not all prey patches are alike; easy-to-swallow prey leads to longer feeding, while tough prey shortens it.

812 - The density of prey patches influences how many sharks show up and how long the frenzy lasts.

813 - Water temperature and current speed influence how odors travel and how quickly a frenzy escalates.

814 - Blood plumes can rise toward the surface, helping surface sharks detect a feeding site.

815 - Communication during feeding is mostly silent and visual; sharks rely on body language and movement.

816 - The term frenzy is a human way to describe rapid feeding; sharks do not experience emotions as humans do.

817 - Movie depictions of synchronized, orderly feeding lines are rare in nature and typically oversimplify behavior.

818 - Real frenzies often involve multiple prey captures in a short time, with fast turnover.

819 - Feeding sites near seal rookeries or fish runs are hotspots for coordinated feeding activity.

820 - Water clarity and prey contrast can affect how readily a shark detects a patch.

821 - Some individuals show tolerance to nearby bites and can move away to avoid injury from a neighbor.

822 - Bite-and-spit behavior helps a shark test whether a piece is edible and safe to swallow.

823 - A shark's bite speed and force depend on its size, health, and motivation during a feeding event.

824 - Turmoil around struggling prey creates disturbance cues that ripple outward to attract more sharks.

825 - Prey that bleeds heavily tends to extend feeding times and attract more participants.

826 - Prey with hard shells or spines can slow digestion and influence how long a frenzy lasts.

827 - Not all sharks participate equally; some stay on the periphery until an opportunity arises.

828 - Baited or discarded food near fishing gear can trigger concentrated feeding events.

829 - A feeding patch can draw other fish and seabirds looking for scraps, creating a multi-species feeding scene.

830 - Hammerheads use their wide heads to corral prey and limit escape routes during a feed.

831 - Great whites may deliver a single decisive bite to immobilize prey before feeding further.

832 - Makos are known for swift, slicing bites that can sever prey's escape attempts.

833 - Sharks often take bites, swallow, and then circle back to nip off more chunks as needed.

834 - Seabirds and other scavengers monitor feeding sites for scraps, sometimes diving in between bites.

835 - Dominant individuals can shift feeding order by signaling or moving aggressively.

836 - When prey density decreases, sharks may spread out or re-locate to new patches.

837 - Chemical cues from fish oils released during blood loss help attract additional sharks.

838 - A feeding patch can be depleted quickly if multiple sharks feed aggressively at once.

839 - Rough seas and adverse weather can disrupt otherwise predictable feeding events.

840 - The bite duration can depend on prey size, toughness, and the shark's hunger level.

841 - Oxygen intake and gill function must keep pace with the high activity during a feeding frenzy.

842 - Sharks may drop or cut a bite if prey proves too large to handle in the moment.

843 - Strong jaw strength enables sharks to deliver deep, secure bites that prevent prey escape.

844 - Tail slaps or sudden bursts of speed can serve as warnings or signals to other sharks.

845 - In some places, boats offload fish guts to study feeding behavior and predator responses.

846 - The presence of divers can cause sharks to pause or slow their feeding, depending on distance and behavior.

847 - Some sharks will feed in close proximity to humans when approached cautiously and with calm behavior.

848 - Gills ventilate water during feeding, helping maintain oxygen flow during high-energy bites.

849 - Sharks monitor each other's position to optimize access to prey while minimizing harm.

850 - Currents and tides move prey patches through feeding grounds, guiding sharks along productive routes.

851 - Prey vulnerability during a frenzy rises with exhaustion, making bites more successful.

852 - Not all bites result in fatal outcomes; many are non-lethal, especially in non-lethal prey interactions.

853 - Prey with tough exoskeletons or spines can cause more damage and prompt faster feeding.

854 - Some sharks tear prey into chunks, while others swallow in larger pieces depending on species.

855 - Sharks use pressure wave cues from surrounding swimmers to gauge how many others are nearby.

856 - Feeding frenzies can help remove weak or injured individuals from prey populations, subtly shaping the ecosystem.

857 - Orcas and other apex predators can influence when and where sharks feed by altering prey availability.

858 - Lunar phase and tides can affect prey movements and the likelihood of a feeding event in some regions.

859 - Feasting hotspots often align with natural prey concentrations, such as schooling fish or seal colonies.

860 - Wave noise and water movement contribute to the acoustic backdrop that sharks sense as feeding cues.

861 - Hunger level directly affects a shark's willingness to participate in a frenzy and take on risk.

862 - Tail-first propulsion can help a feeding shark approach prey with speed and stealth.

863 - The electrical senses in the snout aid prey detection when chemical cues are weak.

864 - A feeding frenzy can attract multiple predator species to a single patch, creating a chaotic but structured scene.

865 - A big bite can provide enough energy to sustain a shark for a long time, influencing its travel decisions.

866 - Juvenile sharks are sometimes drawn to feeding frenzies to learn hunting skills from older individuals.

867 - The scent of blood is a cue but does not guarantee a bite, as prey viability is also weighed.

868 - When prey is too large to swallow in one piece, sharks may stage a multi-step feeding approach.

869 - High prey density can cause sharks to position themselves at different angles around the patch to optimize access.

870 - Sharks adjust their path to minimize energy expenditure while maximizing chances of contact with prey.

871 - After a feeding burst, many sharks enter a rest period to digest and recover.

872 - Bite patterns can be used to identify shark species and infer feeding strategies.

873 - Even during a frenzy, sharks can be selective, favoring certain prey types over others.

874 - Ambient oxygen levels and water quality influence how comfortable sharks are during intense feeding.

875 - Some sharks recover quickly after a bite and immediately re-enter the feeding scene.

876 - Data collected from feeding frenzies helps researchers understand shark behavior and conservation needs.

877 - Feeding opportunities influence how sharks move along coastlines, often guiding their migratory paths.

878 - The first bite often sets the tone for how others will respond to the patch.

879 - Frenzies end when prey resources are exhausted or the patch is fully depleted, leaving the area quiet again.

12

BABY SHARKS! EGGS, PUPS, AND REPRODUCTION THAT GETS WILD

Baby sharks come into the world by two big routes: eggs that hatch outside the body and live births that happen inside the mother. This chapter dives into egg cases, nurseries, and the clever strategies young sharks use to survive.

880 - Mermaid's purses are often anchored to seaweed or rocky surfaces with tendrils.

881 - Inside each egg capsule, the embryo feeds on a yolk sac until it hatches.

882 - Some sharks lay eggs in multiple capsules at once, forming a clutch.

883 - Not all sharks lay eggs; some give birth to live young.

884 - Viviparous pups develop inside the mother and are born alive.

885 - In viviparous species, pups may be nourished by a placenta-like connection.

886 - In other viviparous sharks, embryos receive nourishment via histotroph, a uterus-derived fluid.

887 - Histotroph provides nutrients beyond the yolk for growing embryos.

888 - In some species, embryos also feed on unfertilized eggs produced by the mother in the uterus (oophagy).

889 - Adelphophagy is intrauterine cannibalism where embryos eat siblings.

890 - Sand tiger sharks are well known for adelphophagy, which shapes litter size.

891 - Pups often have color patterns that camouflage them in their nursery habitat.

892 - Newborns are usually capable swimmers and hunters from birth.

893 - Litter sizes vary a lot across species.

894 - Males transfer sperm through claspers during mating.

895 - Courtship can involve chasing, nibbling, and tail-slapping.

896 - Some females can store sperm for months after mating.

897 - Sperm storage lets females time fertilization with favorable conditions.

898 - Eggs or pups can wash up on beaches as mermaid's purses or carcasses.

899 - The shape of mermaid's purses helps scientists identify the species.

900 - Some egg capsules are nearly transparent, so scientists can peek at the developing embryo.

901 - An egg's yolk stores food for the embryo until hatching.

902 - Hatchlings often emerge in shallow, sheltered spots to avoid predators.

903 - Nursery habitats like seagrass beds and mangroves provide food and protection for pups.

904 - The timing of birth often depends on water temperature and prey availability.

905 - Great white sharks are viviparous, and their pups develop in the womb and are born alive.

906 - Hammerhead sharks are viviparous and nourish their developing pups during gestation.

907 - Nurse sharks are viviparous and rely on histotroph to feed developing pups.

908 - Delayed implantation allows fertilized eggs to pause development until conditions are favorable.

909 - Parthenogenesis, reproduction without fertilization, has been observed in some sharks.

910 - Parthenogenetic pups are genetic clones of the mother in those cases.

911 - If a female mates with multiple males, pups can have more than one father in a single litter.

912 - Embryos begin forming sense organs and jaws during gestation.

913 - Egg capsules vary widely in shape, size, and ornamentation across species.

914 - The embryo grows by using yolk and, in some cases, uterine nourishment.

915 - After birth, pups head to nearby bodies of water to hunt.

916 - Nurseries often provide abundant small prey and safety from large predators.

917 - The placenta-like connections in viviparous sharks transfer nutrients to the pup.

918 - The placenta in sharks is different from mammals' placenta but serves a similar function.

919 - Gestation lengths vary widely by species and environment.

920 - The mother's health and food intake influence the development of pups.

921 - Some species reproduce annually; others reproduce less frequently.

922 - Geographical location and ocean conditions influence where sharks breed.

923 - Mating can involve competition among males, including chasing and fighting.

924 - Some populations migrate to specific breeding sites.

925 - The mother's body can undergo changes to support pregnancy.

926 - Some sharks provide little to no parental care after birth; pups must fend for themselves.

927 - The survivability of pups is strongly influenced by habitat quality.

928 - Nurseries are often located near coral reefs, estuaries, and seagrass beds.

929 - The presence of nurseries helps juvenile sharks grow and learn to hunt.

930 - Egg-laying species rely on stable environmental conditions for egg development.

931 - The development of pups is impacted by dissolved oxygen levels.

932 - Newborn sharks quickly learn to avoid threats and search for prey.

933 - The timing of births helps pups face peak prey availability and favorable conditions.

934 - The presence of nurseries is essential for the survival of many species.

935 - Newborn sharks quickly adapt to feeding in their local habitats.

936 - The first weeks of life are the most dangerous for neonates due to predation.

937 - Some populations breed in warm coastal waters while others in cooler regions.

938 - Some eggs hatch in a few months, while others take longer depending on temperature.

939 - Mating and birth events are often seasonal but can occur year-round in some populations.

940 - Parthenogenetic reproduction demonstrates the flexibility of shark life histories.

941 - The fossil record shows mermaid's purses from ancient sharks as evidence of long-standing egg-laying strategies.

942 - Conservation of nursery habitats helps protect future generations of sharks.

943 - Public understanding of shark reproduction supports ocean stewardship.

944 - Shark reproduction is a key part of understanding marine ecosystem dynamics.

945 - Many conservation programs focus on protecting breeding and nursery areas for sharks.

946 - The study of baby sharks reveals the ocean's incredible diversity and life cycles.

13

GROWING UP SHARK: LIFESPANS, MATURITY, AND ANCIENT SURVIVORS

Growing up in the ocean takes a lot longer for sharks than you might think. In this chapter we explore how some sharks grow slowly, how long they live, when they mature, and the ancient lineages that survive to today.

947 - Sharks grow at very different rates, with some species reaching adult size in just a few years while others take decades.

948 - The Greenland shark grows extremely slowly, adding only a small amount of length each year.

949 - Because of slow growth, Greenland sharks often take many decades to reach sexual maturity.

950 - Some Greenland sharks may live for centuries, making them among the longest lived vertebrates.

951 - Scientists estimate shark age using methods such as radiocarbon dating of eye lenses or vertebral bands, depending on the species.

952 - Great white sharks mature much later than many other fish, often not reproducing until their teens or older.

953 - Female great whites generally mature later than males in most populations.

954 - In many shark species, maturity occurs well after birth and depends on growth rates driven by food and habitat.

955 - Whale sharks take several decades to reach reproductive maturity, reflecting their enormous size and slow growth.

956 - Thresher sharks reach maturity after a number of years that varies by species and population.

957 - Sandbar sharks have relatively late maturity compared with small coastal species.

958 - Some sharks such as the epaulette shark can reach reproductive maturity while still small, in shallow tropical waters.

959 - Deep sea sharks often mature more slowly than their shallow water relatives, delaying reproduction to cope with scarce food.

960 - Several shark species are slow breeders, producing only a small number of young each year.

961 - The term living fossil is often used for ancient lineages that survive largely unchanged.

962 - The frilled shark is a classic example of a living fossil, with a body plan that has changed little since the Jurassic.

963 - Goblin sharks and frilled sharks belong to lineages that diverged from other sharks long ago and continue today.

964 - Some modern sharks have skeletons of cartilage, which makes them lighter and more flexible as they grow.

965 - Lanternsharks are small glowing sharks that inhabit deep water and can reach lengths of a meter or two.

966 - Some sharks can survive in rivers after entering freshwater, with bull sharks and river sharks being examples.

967 - Bull sharks have physiological adaptations that let them tolerate freshwater for extended periods.

968 - The walking sharks of the genus Hemiscyllium can move along the seafloor using their pectoral fins as legs.

969 - These walking sharks are found around coral reefs in the Indo-Pacific and can cover short distances by walking between pools.

970 - The bite power of sharks varies widely, and the great white is notable for a strong bite and jaw rotation.

971 - Large sharks such as tiger sharks and great whites deliver powerful bites that have shaped myths about shark feeding behavior.

972 - A typical great white can shed thousands of teeth across a lifetime.

973 - Sharks do not shed scales like skin, but their teeth are replaced constantly by a dental conveyor belt.

974 - The age of Greenland sharks is often cited as a record, though age estimation is complex and ongoing.

975 - The oldest living shark title is disputed because age estimation is challenging in many species.

976 - Some ancient lineages have survived major extinctions due to slow reproduction and wide geographic distribution.

977 - Deep sea environments slow growth but can support long lifespans due to stable conditions and fewer predators.

978 - Some species grow to large sizes yet mature late, creating a long juvenile period.

979 - Age at maturity is influenced by environmental conditions and prey availability.

980 - The idea of sharks being older than dinosaurs reflects the ancient history of sharks as a group.

981 - Sharks as a group existed long before the first dinosaurs roamed the Earth.

982 - The great white lineage has persisted for millions of years as it adapted to various oceans.

983 - The hammerhead lineage diverged early in shark evolution and shows a unique head shape that aids sensing.

984 - The sawshark family demonstrates a different predatory approach with a long, toothed snout.

985 - The sand tiger shark is known for a slow growth pattern and late maturity, affecting its vulnerability to fishing.

986 - Reproductive cycles in sharks can be irregular, with gaps of years between litters in some species.

987 - Gestation in large sharks can last more than a year, longer than many smaller mammals of similar size.

988 - Pups per litter vary greatly between species, from a handful to more than a dozen.

989 - High juvenile mortality in the wild means long lifespans help populations persist over time.

990 - Genetic studies show that populations of the same

species can mature at different rates depending on local conditions.

991 - Greenland sharks grow slowly and their growth rate remains one of the least understood aspects of large predators.

992 - Some sharks can inhabit low oxygen environments, which may aid longevity by reducing metabolic stress in some habitats.

993 - Longevity in sharks is often linked to large body size and slow metabolism.

994 - The aging patterns of sharks are not as well understood as in mammals, with some species showing unusual aging dynamics.

995 - Late life reproduction in some long lived sharks helps maintain populations in changing oceans.

996 - For many sharks, surviving through many years with steady reproduction is more important than rapid breeding.

997 - The fossil record shows sharks have endured major climate shifts by adapting to changing seas and prey.

998 - Some marine lineages have endured for tens of millions of years, including several ancient sharks.

999 - The longevity of long lived sharks can make them important indicators of past environmental changes.

1000 - In the wild, shark populations often show more young individuals due to fishing pressure, affecting observed lifespans.

1001 - Conservation plans stress protecting adult sharks to maintain reproductive output and ecosystem roles.

1002 - The spiny dogfish is known for a long lifespan and slow growth.

1003 - Spiny dogfish can live for several decades in cooler waters.

1004 - The roughskin dogfish is another example of a slow growing, long lived species.

1005 - The vastness of the ocean helps long lived sharks avoid predators and persist across oceans.

1006 - Walking sharks are unusual athletes among sharks, though longevity in this group is not exceptionally different from related species.

1007 - Sharks reproduce in seasonal pulses in some species, while others breed year round depending on food and water conditions.

1008 - Longevity in sharks supports their role as apex predators for many decades.

1009 - Understanding age structure in sharks helps scientists predict how populations recover after declines.

1010 - The frilled shark's ancient lineage adds depth to our understanding of marine evolution.

1011 - Fossil relatives show close links to modern lines, including the frilled shark.

1012 - Long lived sharks can pass on genetic material across millions of years of evolution.

1013 - In some species, females produce fewer pups but invest in larger offspring that survive longer.

1014 - The distribution of old and young individuals in populations reveals how past environmental changes shaped life histories.

1015 - The ability to endure lean times in deep water helps some sharks live longer lives.

1016 - Deep water habitats can protect sharks from some threats that affect shallow water populations.

1017 - Spawning migrations are influenced by temperature and ocean cycles, affecting when sharks mature and reproduce.

1018 - Growth rates in sharks are tightly linked to prey availability and metabolism.

1019 - A shark total lifespan includes juvenile and adult phases, not just the reproductive years.

1020 - Some ancient shark genera persisted for tens of millions of years with little shape change.

1021 - Deep sea ecosystems have produced many long lived sharks adapted to cold, dark conditions.

1022 - The frilled shark has a distinctive face and fins that evoke a prehistoric look.

1023 - Ancient shark lineages offer clues about how life in the oceans responded to mass extinctions.

1024 - Some species migrate across oceans for many years, contributing to long lived populations.

1025 - The ability to detect electric fields helps many sharks find prey even when hunting is difficult, supporting longevity.

1026 - The living fossil label describes sharks that remain similar to ancient relatives, though evolution continues.

1027 - Shark life histories range from fast growing and short lived to slow growing and long lasting.

1028 - Some sharks have staggered breeding cycles that help populations persist despite environmental fluctuations.

1029 - The ability to thrive across a range of habitats has helped some ancient sharks endure climate changes.

1030 - Animal aging studies show that reading age in sharks relies on biological methods other than tree ring analogies.

1031 - Long lived sharks provide steady ecological roles and preserve genetic lineages over long periods.

1032 - Understanding how sharks grow, age, and mature helps scientists predict threats and protect these living legends.

14

SHARKS ON THE MAP: WHERE THEY LIVE AND HOW THEY TRAVEL

Sharks travel a hidden map beneath the waves, from coral reefs to the far open ocean. They use scent, electricity, and the pull of the planet's magnetic field to roam across vast distances. In this chapter, you'll learn where sharks live, how they move, and the currents that guide their journeys.

1033 - Many species begin life in shallow coastal nurseries where warm, murky water helps shield young sharks from predators.

1034 - Coral and rocky reefs provide abundant prey and complex cover that shape where many sharks live day to day.

1035 - The open ocean hosts some of the most expansive shark habitats, where millions of gallons of water carry food and signals for travelers.

1036 - Some sharks regularly venture into brackish or even river mouths, especially if prey items drift into the estuary.

1037 - Acoustic and satellite tags have mapped migrations that can stretch across entire oceans and back again.

1038 - Ocean upwelling zones, where nutrient-rich water rises, attract schools and the sharks that chase them.

1039 - Ice edge regions provide feeding opportunities for species that can tolerate cold water, like certain sharks found near Greenland and Antarctica.

1040 - Reefs, seamounts, and continental shelves form predictable stages where sharks hunt and rest between movements.

1041 - Habitat use varies widely by species, with some preferring warm surface waters and others cruising at depth.

1042 - Great white sharks often migrate along coastlines, riding currents and following seals and other prey.

1043 - Hammerhead sharks are famous for roaming near reefs and island chains where prey schools are abundant.

1044 - Tiger sharks roam across open seas and tropical coasts, following a broad range of prey animals.

1045 - Silky sharks travel in deep water on the edge of the continental shelf, often in large groups.

1046 - Whale sharks migrate long distances to feed on dense plankton blooms in warm waters.

1047 - Thresher sharks use their long tails to herd prey in open waters near the surface or along thermoclines.

1048 - Some sharks exhibit diel patterns, moving to different depths between day and night to hunt.

1049 - Seasonal migrations are common in temperate zones as water temperature changes and prey moves.

1050 - Satellite tagging has shown that some sharks can traverse entire ocean basins, crossing barriers like unfavorable currents.

1051 - Genetic studies reveal population structure that maps where different shark populations primarily live and breed.

1052 - Sharks possess a remarkable ability to detect magnetic fields, which scientists suspect helps with large-scale navigation.

1053 - Acoustic arrays placed on coastlines and seafloor help researchers track sharks as they move through given channels.

1054 - Some sharks use scent trails left by prey as a guide to follow feeding grounds across hundreds of kilometers.

1055 - Long currents like the Gulf Stream act as natural

highways for many migratory sharks along the U.S. east coast.

1056 - The East Australian Current helps some species move gracefully along the eastern coast of Australia to new feeding zones.

1057 - Oceanic fronts, where warm and cold waters meet, often concentrate prey and attract traveling sharks.

1058 - The Antarctic and sub-Antarctic oceans host species that travel long distances around the Southern Ocean to find food.

1059 - Some sharks undertake transoceanic journeys that last many months and span thousands of miles.

1060 - Juvenile sharks often stay closer to shore, gradually venturing farther as they grow and learn.

1061 - Sharks use high-contrast visual cues and lighting to spot prey and obstacles during long swims.

1062 - The breeding season can influence migration timing, with some females traveling to specific nurseries to give birth.

1063 - Acoustic transmitters provide real-time location data that helps scientists map migration corridors.

1064 - Tagging programs have shown that juvenile tiger sharks can cover hundreds of kilometers within a single year.

1065 - Some sharks use seismic or acoustic ripples in water to sense approaching prey schools.

1066 - Drag and hydrodynamics play a role in how far and how quickly sharks can migrate in different environments.

1067 - Riverine and estuarine species are especially affected by seasonal rainfall and river discharge patterns.

1068 - Satellite data mirrors climate patterns, helping researchers predict when sharks will move toward feeding grounds.

1069 - Not all migrations are for feeding; some are tied to mating habitats and genetic diversity.

1070 - Some sharks ride seasonal wind-driven upwelling to extend their travel without expending excessive energy.

1071 - Ocean highways are not straight lines; they curve with currents, eddies, and geography like coastlines and islands.

1072 - Some populations are isolated by land barriers and have distinct migratory routes and timing.

1073 - Acoustic arrays can detect even small-scale pauses or changes in a shark's travel pattern.

1074 - Sharks sometimes rest by gliding in the upper layers of water while drifting with currents.

1075 - The seasonal ice cover in the polar regions creates seasonal migration windows for some species.

1076 - Tracking data has revealed that many sharks make detours to exploit predictable feeding opportunities.

1077 - Tidal cycles can influence the timing of shallow-water migrations along coastlines.

1078 - The color and shape of the sea floor in a region can influence where sharks hunt and rest.

1079 - Some species prefer to follow the edges of continental shelves where prey density is high.

1080 - In some places, sharks ride the same currents year after year, forming repeat migration routes.

1081 - Migration routes are influenced by prey distribution as much as by temperature or salinity.

1082 - Some sharks reach reproductive or nursery grounds that are biologically protected by currents that keep predators away.

1083 - Young sharks in nurseries learn to navigate by following adults or learned chemical cues from prey.

1084 - The presence of quieter inland seas or lagoons can serve as safe spawning zones for some species.

1085 - Ocean noise pollution can alter migration patterns by affecting prey behavior and predator detection.

1086 - Changes in sea surface temperature influence where sharks will go for feeding as fish distributions shift.

1087 - Some sharks undertake annual migrations that mirror the life cycles of their prey, like tuna or sardines.

1088 - The combined use of satellite and acoustic technology provides more accurate maps of migratory corridors.

1089 - Sharks near archipelagos often use the geometry of island chains to steer around dangerous currents.

1090 - Climate change is shifting traditional migration routes for some species as oceans warm and currents alter.

1091 - Some sharks demonstrate leap-like bursts during migration when chasing a fast-moving prey school.

1092 - The concept of "navigation by magnetism" remains a topic of active study in shark biology.

1093 - Some migratory routes run along the margins of sea ice, offering feeding opportunities and shelter.

1094 - Sharks can sense the electrical signaling of neurons in prey, guiding them toward feeding hotspots.

1095 - Tag data shows that some sharks cross and recross the same coastlines year after year.

1096 - The longest known migrations for some species stretch over thousands of kilometers in a single journey.

1097 - Ocean temperature gradients often act as invisible signposts for migratory sharks.

1098 - Estuaries with rich prey can act as magnets for young sharks stepping into adulthood.

1099 - The distribution of prey is often patchy, causing sharks to alter their routes to follow blooms.

1100 - Some sharks return to the same reef system to mate because those reefs provide both shelter and food.

1101 - Researchers often rely on a combination of tracking methods to overcome data gaps.

1102 - Magnetic navigation in sharks is thought to be learned from juvenile experience and refined with age.

1103 - The Great Barrier Reef region hosts multiple migratory corridors used by several shark species.

1104 - Some sharks travel along coastlines but switch offshore for periods to exploit deep-water prey.

1105 - There is evidence that sharks can detect changes in water chemistry linked to prey migrations.

1106 - In some areas, seasonal "feeding storms" attract dense aggregations of prey and chasing sharks.

1107 - Tag returns from some islands reveal that sharks can travel back-and-forth between far-flung sites.

1108 - Sharks often use bathymetric features like ridges and trenches to navigate and hunt.

1109 - Juvenile sharks may follow their mothers' how-to navigate paths before striking out alone.

1110 - Some migrations are timed to coincide with spawning migrations of other animals, like humpback whales.

1111 - In the deepest oceans, some sharks rely on stable temperature layers to minimize energy use during long swims.

1112 - In warm-tropical zones, sharks hug the surface more frequently to take advantage of air-breathing currents called upwellings.

1113 - Some species migrate through the central Pacific along long, high-latitude corridors.

1114 - Tag data reveals that some sharks can stay within the same large gyre for extended periods.

1115 - The term "river shark" refers to a species that is adapted to freshwater but is uncommon and often misunderstood.

1116 - The presence of mangroves provides rich feeding for juvenile sharks while they learn migration basics.

1117 - Some sharks display remarkable endurance, swimming for days at a time to reach a preferred feeding zone.

1118 - Mercury contamination in some coastal waters can indirectly affect sharks by altering prey behavior and distribution.

1119 - Conservation research emphasizes protecting migratory corridors to safeguard genetic diversity.

1120 - Not all sharks migrate; some populations are resident and stay in one area for life.

1121 - Scientists use drone imagery to observe surface movement patterns that may indicate migration.

1122 - Ocean highways are often dynamic, changing year to year with climate shifts and prey availability.

1123 - Some trackers have shown sharks moving in looping patterns, suggesting exploration as well as migration.

1124 - The Arctic hosts some of the most enigmatic shark migrations, with species adapting to extreme cold.

1125 - The deep Gulf of Mexico hosts migrations tied to prey such as squid and cuttlefish that gather in certain seasons.

1126 - Some sharks can travel thousands of kilometers along the margins of continents rather than straight across oceans.

1127 - Portable cameras attached to sharks have given

scientists a unique first-person view of long-distance travel.

1128 - Not every journey ends at a food source; some migrations help maintain genetic connectivity between populations.

1129 - The study of shark migration combines oceanography, biology, and data science to reveal hidden routes.

1130 - Even without perfect maps, scientists can predict likely migration corridors by combining prey surveys, temperature data, and observed movements.

15

FRESHWATER AND RIVER SHARKS: WHEN SALTWATER RULES DON'T APPLY

Some sharks aren't content to stay in the ocean. They swim into rivers and lakes, proving that saltwater isn't the only home for these toothy travelers. This chapter dives into the surprising world of freshwater and river sharks and how they survive when salt rules don't apply.

1131 - Bull sharks are the best-known freshwater travelers and are famous for entering rivers far from the coast.

1132 - They can tolerate a wide range of salinity, from near freshwater to saltwater.

1133 - In seawater, bull sharks maintain high levels of urea in their blood to stay isotonic with the surrounding water.

1134 - When they move into rivers, their kidneys adjust to excrete more dilute urine and conserve salts.

1135 - Bull sharks have been documented hundreds of miles inland in large river systems like the Mississippi and Amazon.

1136 - The Ganges shark, Glyphis gangeticus, is one of the few sharks believed to spend significant time in freshwater in the Ganges and Brahmaputra basins.

1137 - The speartooth shark, Glyphis glyphis, inhabits riverine habitats in northern Australia and nearby estuaries.

1138 - Glyphis species are among the rarest and most elusive sharks because they spend much of their lives in murky rivers and are hard to study.

1139 - River sharks often rely on non-visual senses in turbid waters, rather than sight alone.

1140 - Their sense of smell is extremely acute, helping them locate prey in muddy rivers.

1141 - River-dwelling sharks feed on fish, crustaceans, and sometimes birds or small mammals that fall into water.

1142 - Bull sharks tend to have a broad snout that helps detect vibrations in the water.

1143 - Their teeth are sharp and serrated, suited for gripping slippery prey and tearing flesh.

1144 - Bull sharks can reach lengths of around 11 feet (3.4 meters).

1145 - Many river sharks give birth to live young, with pups nourished by a placental connection.

1146 - Gestation periods for large river sharks are lengthy, often near a year.

1147 - Juvenile river sharks often inhabit brackish nursery areas before venturing into deeper rivers.

1148 - Some sharks that enter freshwater are not permanent specialists; they use rivers for feeding and nurseries.

1149 - The term freshwater shark is a simplification; many of these species move between freshwater and brackish zones.

1150 - The Ganges shark is endangered by habitat loss, pollution, and fishing pressure.

1151 - Dams and water management can fragment river habitats and block shark migrations.

1152 - Bycatch in fisheries threatens Glyphis species and other river sharks.

1153 - River sharks typically have slow population growth, making recovery slow after declines.

1154 - Because they inhabit remote river systems, many river sharks remain poorly understood by science.

1155 - Scientists use acoustic tags and telemetry to study the movements of river sharks.

1156 - Flood pulses during the rainy season create new feeding opportunities in river floodplains.

1157 - River sharks may migrate up rivers with seasonal floods to exploit shifting prey landscapes.

1158 - Some river sharks can tolerate brief periods of low oxygen in floodwaters.

1159 - River sharks rely on a combination of senses: smell, electroreception, and hearing, to locate prey in murky water.

1160 - These sharks challenge the stereotype that all sharks live only in the ocean.

1161 - Glyphis represents one of the most enigmatic and ancient lineages among living sharks, adapted to freshwater life.

1162 - River sharks have adapted to freshwater life without losing their identity as true sharks.

1163 - River sharks are more commonly found in tropical and subtropical regions where rivers meet the sea.

1164 - The Ganges River system hosts slow-moving stretches and oxbow lakes that attract river sharks.

1165 - Freshwater and brackish zones provide abundant prey like small fish and crustaceans for river sharks.

1166 - River sharks often hunt along river banks where prey congregates during floods.

1167 - The ability to survive gradients of salinity relies on specialized kidney function and ion transporters.

1168 - Some river sharks exhibit site fidelity to particular river sections, while others roam more widely.

1169 - The flow and sediment of rivers influence prey availability and habitat structure for sharks.

1170 - Their teeth can wear from feeding on hard-shelled prey like crustaceans.

1171 - River sharks often show patience in murky water, waiting for prey to pass by.

1172 - In some regions, river sharks are shy and avoid boats and humans.

1173 - Glyphis species typically have small litter sizes and slow growth, which hampers rapid population recovery.

1174 - The Ganges shark remains one of the most poorly understood sharks due to limited sightings.

1175 - Local communities sometimes feature river sharks in folklore and cautionary tales.

1176 - Protecting river habitats benefits many species that depend on these freshwater ecosystems.

1177 - True freshwater residency is rare for most sharks; many true river-dwelling species live in brackish water most of the time.

1178 - River sharks play roles in controlling prey populations and helping keep river ecosystems balanced.

1179 - The diets of river sharks are predominantly fish-based, with occasional crustaceans.

1180 - Some river sharks have been observed leaping briefly out of the water during feeding or courtship displays.

1181 - Scientists warn that many river sharks are data-deficient due to the difficulty of studying them in remote habitats.

1182 - River-dwelling sharks can show remarkable tolerance for changing environmental conditions.

1183 - Conservationists emphasize the importance of preserving river connectivity to protect these sharks.

1184 - River sharks are often caught as bycatch in gillnets and longlines set in river mouths.

1185 - The life histories of Glyphis species feature relatively slow growth and late maturity.

1186 - River sharks depend on clear water in some areas to support ambush hunting, even as many are found in murky habitats.

1187 - Some river sharks have well-developed lateral lines to detect water movement.

1188 - The distribution of Glyphis species is highly fragmented, with confirmed sightings in only a few river systems.

1189 - The word 'river shark' can be used to describe multiple, distantly related species that share the river-dwelling lifestyle.

1190 - Many river sharks use estuaries as transition zones for moving between river and sea environments.

1191 - Tagging studies in some regions show river sharks regularly migrate along key river corridors.

1192 - River sharks contribute to the biodiversity value of freshwater systems and attract ecotourism.

1193 - The teeth of river sharks are generally sharp and designed for grabbing slippery prey.

1194 - Some river sharks show rapid bursts of speed to ambush prey in tight spaces.

1195 - River flood cycles can lead to seasonal spikes in prey density that support shark growth.

1196 - The ecology of river sharks is influenced by human land use, agriculture, and pollution.

1197 - River sharks may shelter under submerged logs or in complex root systems along river banks.

1198 - Members of Glyphis and related river-dwelling sharks illustrate how sharks adapt to freshwater environments without sacrificing their predatory style.

1199 - The ability to navigate both salt and freshwater is a testament to the plasticity of shark physiology.

1200 - Public interest in river sharks has grown thanks to documentaries and educational programs.

1201 - The river environment can offer safety from coastal storms and large predators for juvenile sharks.

1202 - River sharks may display crepuscular activity, hunting around dawn and dusk.

1203 - Some river sharks feed during the night when water visibility is lowest.

1204 - River sharks can be affected by overfishing, habitat degradation, and climate change.

1205 - Sustainable fisheries and habitat protection are key to safeguarding river sharks.

1206 - River sharks underscore the interconnectedness of marine and freshwater ecosystems.

1207 - The study of river sharks helps scientists understand how osmoregulation evolved in cartilaginous fishes.

1208 - Some river sharks contribute to local economies through eco-tourism and shark-watching activities.

1209 - The presence of river sharks can indicate good water quality and healthy predator-prey dynamics in a river.

1210 - Some documentaries highlight river sharks for their startling ability to move between ecosystems.

1211 - River sharks are a reminder that not all apex predators stay in the open ocean.

1212 - In aquariums, keeping river sharks requires recreating brackish conditions and ample hiding spaces.

1213 - Glyphis species are primarily found in riverine and estuarine habitats in Asia and Australia.

1214 - The Ganges River is home to a unique diversity of aquatic life, including river sharks.

1215 - River sharks face threats from habitat fragmentation caused by damming and pollution.

1216 - Local communities can participate in conservation by reporting sightings and protecting nesting areas.

1217 - Research into river sharks continues to shed light on their behavior, movement, and ecology.

1218 - River sharks can tolerate seasonal variations in water temperature and salinity.

1219 - The life cycle of river sharks includes juvenile stages that rely on nursery habitats for survival.

1220 - The ecological role of river sharks includes linking river and sea food webs through migrations.

1221 - River sharks illustrate the amazing range of habitats that sharks can occupy.

1222 - Public education about river sharks helps dispel myths and promotes conservation.

1223 - The study of river sharks integrates biology, ecology, hydrology, and conservation science.

1224 - Many river sharks are more at risk from human activity than from other predators.

1225 - Protecting river corridors benefits many species in addition to sharks.

1226 - River sharks demonstrate that nature's ocean creatures can thrive in freshwater environments too.

16

TEAMWORK, ATTITUDE, AND SOCIAL LIVES: SHARK BEHAVIOR UP CLOSE

Sharks aren't lone predators; many species display surprising social lives, bold attitudes, and even teamwork. In this chapter, we explore how schooling, solo hunting, dominance displays, curiosity, rest, and silent communication shape shark behavior up close.

1227 - Many sharks form schools to boost hunting power and increase predator awareness, especially in species that feed on schooling prey.

1228 - The wide, flat hammerhead head helps these sharks spot prey from multiple directions, aiding coordinated hunting.

1229 - Blacktip reef sharks often gather around cleaning stations, creating small social gatherings as they interact with cleaners and other sharks.

1230 - Lemon sharks in warm coastal habitats display stable social networks, with some individuals preferring to associate with certain partners.

1231 - Reef sharks can show dominance displays at feeding sites, signaling who will get access to prey first.

1232 - Dominance signals include upright posture, slow deliberate movements, and raised pectoral fins during confrontations.

1233 - A sharp jaw snap or quick tail slap can serve as a warning to rivals without escalating into a fight.

1234 - When meeting unfamiliar sharks, individuals may circle each other cautiously as a form of social introduction.

1235 - The lateral line system lets sharks sense nearby movement, helping them coordinate during group movements.

1236 - Olfactory cues help sharks locate prey and may play a role in following conspecifics to a food source.

1237 - Cleaning stations are social hubs where sharks routinely interact with cleaners and with each other during parasite removal.

1238 - Some species perform quiet greeting sequences, such as slow circling and gentle body nudges, when re-encountering known partners.

1239 - Sharks rely on their electroreception to detect electrical fields, which can help them sense other sharks in murky water.

1240 - Sharks do not vocalize; most of their communication is visual, tactile, or via swimming patterns.

1241 - Resting nurse sharks can stay motionless for long periods, especially in safe, low-predation zones.

1242 - In some species, individuals maintain a low-activity state during rest, yet remain capable of quick responses if threatened.

1243 - Sentry or lookout behavior has been observed in some reef sharks, with one shark watching while others feed.

1244 - Sharing space around a food patch is common when prey is abundant, with less aggression than expected.

1245 - Hammerhead schools can help corral schools of prey, making it easier for the group to feed.

1246 - Gray reef sharks sometimes form loose daytime groups near reef passes where prey aggregations occur.

1247 - Under certain conditions, different shark species may share the same reef area without intense conflict when resources allow.

1248 - Young sharks often stay close to larger, experienced individuals during early life in many species to learn safe behavior.

1249 - Curiosity drives sharks to approach new objects or divers with cautious inspection, rather than immediate aggression.

1250 - If a non-threatening shark approaches, others may display mild avoidance rather than escalation.

1251 - When crowded, sharks adjust spacing to avoid collision and maintain fluid movement.

1252 - Individual sharks can show personalities, with some individuals consistently bolder or more exploratory than others.

1253 - Social tendencies in sharks can change with age, season, and local prey availability.

1254 - The level of social interaction in sharks is highly species-dependent; what is true for hammerheads may not hold for great whites.

1255 - Reef sharks often display greeting rituals with familiar neighbors including measured circling and proximity checks.

1256 - Nonverbal communication is central to shark social life due to their lack of vocalizations.

1257 - Some studies have documented recurring social associations among lemon sharks across weeks in particular lagoons.

1258 - Social clustering near key reef features can offer protection and access to partner pairings.

1259 - Sharks can form longer-term associations with specific individuals, and these bonds can influence movement patterns.

1260 - The presence of large predators can reshape social networks by pushing individuals to cluster or disperse.

1261 - When prey is plentiful, competitive aggression often decreases, allowing calmer social interactions.

1262 - Social networks in sharks can be studied by tracking close associations over time, revealing preferred partners.

1263 - The approach orientation of two sharks, whether head-on or from the side, conveys different social meanings.

1264 - Subtle postural cues, such as fin positioning, help signal intent during social encounters.

1265 - Daytime and nighttime patterns can differ in social activity depending on species and habitat.

1266 - Socially aware sharks may adjust their hunting tactics in the presence of conspecifics to avoid overlap.

1267 - A shark's personality can influence how often it initiates social contact at feeding spots.

1268 - Some sharks appear to greet trusted partners with closer swimming proximity and slower movements.

1269 - Social interactions around a shared resource can reduce aggression and promote efficient feeding.

1270 - Younger sharks often rely on peers to learn the location of fruitful feeding grounds.

1271 - Cleaner stations can be stepping stones for social contact and tolerance between different sharks.

1272 - Sharks may show tolerance toward others near a safe, resource-rich environment like a reef cleaning station.

1273 - Using multiple senses, sharks gather information about social environment and potential partners.

1274 - Observational learning lets juveniles imitate successful foraging techniques.

1275 - Social behavior around prey patches is shaped by resource density, competition, risk.

1276 - Some individuals demonstrate more willingness to approach divers or unfamiliar underwater objects.

1277 - Sharks may adjust their speed and trajectory during group movement to avoid collisions.

1278 - The distribution of prey can influence how long a group remains together in a general area.

1279 - Sharks may approach slowly and cautiously toward a new group, testing reaction before joining.

1280 - The social life of sharks contributes to reef ecosystem dynamics by coordinating movement and feeding.

1281 - Mutualistic cleaning interactions are a critical part of social dynamics around reef sharks.

1282 - The difference between school and shoal can reflect degree of organization in social groups.

1283 - Some sharks show systematic avoidance of crowded spaces with unfamiliar individuals.

1284 - Reef environments create opportunities for repeated social encounters among same peers over time.

1285 - Sensing through multiple senses supports coordinated movement in social settings.

1286 - Sharks' social behavior helps exploit ephemeral prey blooms by attending them as a group.

1287 - Long-term studies in lagoons have shown stable assortative mixing among certain shark individuals.

1288 - Proximity data reveals that sharks travel with familiar companions across days or weeks.

1289 - Cleaning stations reduce aggression as they learn routines and expectations with cleaners and other sharks.

1290 - The social life of sharks is a key to understanding how predators cope with crowded reef habitats.

1291 - Some sharks adjust social approach depending on whether they are alone or with juveniles.

1292 - Young sharks often travel with peers of the same age class, forming juvenile groups.

1293 - The social behavior of sharks changes with breeding seasons, reflecting shifts in resource use.

1294 - The hidden world of shark social life is revealed through careful observation, tagging, and underwater video.

1295 - The silent communication among sharks relies on timing and precise body language to avoid conflict.

1296 - Some sharks display tolerance toward others at feeding sites even when competition exists.

1297 - The social life of sharks shows a spectrum from solitary hunters to highly social groups.

1298 - Cleaners and clients create a dynamic social network at reef systems with frequent interactions.

1299 - Shared habitats can foster long-term associations when individuals repeatedly encounter the same resources.

1300 - Social interactions influence how sharks use habitat features during foraging.

1301 - Divers can observe notable social behavior around shark schools as they feed.

1302 - The study of shark social behavior helps scientists understand how populations adapt to changing oceans.

1303 - Social cues can influence mate-choice indirectly, through resource access, even when mating isn't visible.

1304 - Sharks may show consistent preferences for familiar divers who have previously interacted with them.

1305 - The broad variety of shark social behavior offers insight into predator communication and organization.

1306 - Even solitary species may rely on social information to locate prey or safe resting spots.

1307 - The ways sharks communicate through movements are a form of language scientists are decoding.

1308 - The combination of senses helps sharks navigate crowded social scenes in complex reefs.

1309 - The social dynamics of sharks can modulate ecosystem-level processes like prey distribution.

1310 - The presence of robust social groups can influence reef recovery after disturbances.

1311 - Shark social behavior is dynamic and context-dependent, changing with prey availability and habitat structure.

1312 - Their social life includes peaceful co-existence rather than constant aggression.

1313 - Understanding shark social behavior enhances appreciation for intelligence and adaptability.

1314 - Some sharks show tolerance around non-threatening conspecifics.

1315 - The social world of sharks is a frontier of marine biology.

1316 - Observations around reefs show calm, exploratory social interactions.

1317 - Diet-related social behavior includes cooperation around abundant resources.

1318 - Diurnal and nocturnal variations add complexity to social activity.

1319 - Movement is influenced by currents, prey, and social ties.

1320 - Studying shark social life helps conservation and education.

17

MYTH-BUSTERS: WHAT MOVIES AND LEGENDS GOT TOTALLY WRONG

Movies and legends love to paint sharks as unstoppable man-eaters. In this chapter, we separate splash from science and reveal the truths behind famous shark myths.

1321 - The vast majority of shark species pose no threat to humans, and most interactions between people and sharks end without a bite.

1322 - Only a small handful of shark species account for most unprovoked bites on people, not every shark is a danger.

1323 - When bites happen, they're usually exploratory or defensive rather than a calculated hunt for humans.

1324 - The term man-eater is a sensational label used by media; scientists categorize sharks by species and behavior, not by human targets.

1325 - Jaws and similar films amplified fear by depicting a single, beach-haunting predator; in the real oceans sharks hunt a variety of prey and patterns vary widely.

1326 - Sharks are not mindless killing machines; they have complex behaviors, problem-solving abilities, and nuanced hunting strategies.

1327 - The great white's bite is incredibly powerful, with estimates around 1.8–2 tons of force.

1328 - A single bite is not always fatal by itself; outcomes depend on injury location, depth, and medical response.

1329 - The idea that a tiny drop of blood will attract sharks from miles away is an oversimplification; scent dispersal depends on currents, depth, and other cues.

1330 - Bioluminescence in sharks is typically used for camouflage or signaling, not to lure humans.

1331 - The cookiecutter shark bites round, plug-shaped chunks out of larger animals, a distinctive feeding strategy.

1332 - Shark reproduction comes in several modes: oviparity (egg-laying), viviparity (live birth with placenta), and ovoviviparity (eggs hatch inside the mother).

HUMANS VS. SHARKS: CONSERVATION, THREATS, AND COMEBACKS

Humans and sharks share the oceans, but human actions have put many shark species at risk. This chapter explores the threats sharks face from fishing, habitat loss, and climate change, and it highlights the conservation wins that show recovery is possible. Readers will learn practical ways to help sharks survive and thrive.

1333 - Humans fish for sharks for meat and fins, driving declines in many species.

1334 - Finning, removing fins and discarding the body, is banned or restricted in many places but still occurs illegally.

1335 - Bycatch in longlines, nets, and other gear kills thousands of sharks that are not the target catch.

1336 - Some shark species have lifespans of multiple decades, further slowing rebounds after depletion.

1337 - Overfishing reduces the pool of breeding adults, shrinking future populations.

1338 - Habitat loss, including mangroves and seagrass beds, erodes nursery areas for young sharks.

1339 - Coastal development and pollution degrade critical habitats and prey bases for sharks.

1340 - Climate change alters ocean temperatures and currents, shifting where sharks feed and breed.

1341 - Ocean acidification can affect prey species, influencing the food web sharks depend on.

1342 - Overfishing of predator prey can indirectly affect sharks by changing prey availability and ecosystem balance.

1343 - Some regions have declared shark sanctuaries or implemented seasonal bans to reduce fishing pressure.

1344 - Marine protected areas help sharks by safeguarding essential habitats like nurseries and breeding sites.

1345 - Catch-and-release practices can reduce mortality when performed carefully and quickly.

1346 - Bycatch reduction devices and gear changes can dramatically cut shark deaths in fisheries.

1347 - International trade controls under CITES help regulate shark products to prevent overexploitation.

1348 - Whale sharks receive protection in many countries and international agreements due to slow reproduction and population declines.

1349 - Whale sharks are listed on CITES Appendix II, requiring permits for international trade.

1350 - Hammerhead sharks, especially the great hammerhead, have faced severe declines prompting protective measures.

1351 - Silky sharks have suffered declines in some regions due to bycatch and fins for sale.

1352 - Blue sharks are widespread but show regional declines from fishing pressure.

1353 - Some sharks, like mako, are prized for speed and fins, driving targeting in some fisheries.

1354 - Many sharks migrate across ocean basins, requiring international cooperation to protect migratory populations.

1355 - Sharks rely on stealth, speed, and powerful jaws to catch prey, often ambushing from below.

1356 - Sharks detect vibrations and chemical cues in water to locate prey and navigate the ocean.

1357 - Juveniles often rely on shallow nursery habitats like mangroves and seagrass to avoid predators.

1358 - Nearshore waters and popular beaches can bring sharks into contact with humans, underscoring the need for safety and awareness.

1359 - Sharks are ecological indicators of healthy marine ecosystems; their presence often signals productive habitats.

1360 - Removing apex predators like sharks can trigger trophic cascades that reshape entire ecosystems.

1361 - Some shark species are more resilient to fishing pressure than others because of different reproductive strategies.

1362 - With protections in place, some shark populations can recover, though timelines vary by species and region.

1363 - Large protected areas and sanctuaries help migratory sharks access safe habitats across oceans.

1364 - Protecting sharks also helps maintain ecosystem services that support fisheries and reef health.

1365 - Responsible ecotourism, including shark diving, can provide economic incentives to protect sharks.

1366 - Community-based conservation can succeed when local fishers gain long-term benefits from healthy shark populations.

1367 - Some governments require bycatch reporting and data sharing to improve management.

1368 - Non-government organizations play a key role in translating science into policy and on-the-ground protection.

1369 - The decline of sharks can influence the abundance of mid-level predators and prey, altering food webs.

1370 - Deep-sea shark species face unique threats from longline fishing and high bycatch rates in some regions.

1371 - Regular scientific surveys track shark abundance and help adjust management strategies.

1372 - Local communities often hold traditional knowledge that can complement formal protections for sharks.

1373 - Public education about sharks' ecological roles helps reduce fear and build support for conservation.

1374 - The health of shark populations often reflects the overall condition of marine ecosystems.

1375 - Many shark species undertake seasonal movements tied to prey availability, breeding, or temperature.

1376 - Modern tracking technology, including satellite tags, reveals surprising migratory routes of many sharks.

1377 - Some sharks show fidelity to particular feeding grounds, returning to the same areas year after year.

1378 - Some fisheries restrict landing of young sharks to protect juvenile cohorts and promote recovery.

1379 - No-take zones, where fishing is prohibited, exist in some regions to protect sharks.

1380 - Sharks can serve as indicators of water quality, since pollution can affect their prey and health.

1381 - Protected areas often overlap with zones important for other endangered marine species, maximizing conservation benefits.

1382 - International cooperation is essential because many sharks cross multiple national jurisdictions.

1383 - Climate-induced changes in currents can affect larval survival and distribution of sharks.

1384 - Mercury and other toxins can accumulate in shark flesh, leading to consumer health considerations.

1385 - Historically, shark liver oil and other byproducts attracted hunting pressure, contributing to declines.

1386 - The economic value of sharks extends to tourism, sport fishing, and fisheries, making conservation economically relevant.

1387 - Licensing schemes for shark fishing help ensure accountability and sustainable harvests.

1388 - Shark sanctuaries in places like Palau and the Bahamas reduce illegal fishing and support recovery.

1389 - In some regions, protected measures have led to increases in certain shark populations.

1390 - Data on sharks are challenging to collect due to wide ranges and elusive behavior.

1391 - Citizen science and beach surveys help track shark sightings and trends over time.

1392 - Some bycatch programs provide incentives to release sharks alive and avoid unnecessary harm.

1393 - Sustainable shark-fishing certifications help direct demand toward responsibly managed products.

1394 - Healthy reefs with abundant herbivores support prey communities for sharks and improve overall resilience.

1395 - Climate refugia are areas that remain suitable for sharks even as conditions shift elsewhere.

1396 - Marine spatial planning helps designate critical shark habitats and regulates fishing around them.

1397 - The global distribution of sharks is shifting, with some regions seeing increases while others decline.

1398 - Shark-watching tourism can offer a profitable alternative to harvesting in some communities.

1399 - Recovery stories exist for species like certain hammerheads in protected areas, demonstrating potential for rebound.

1400 - The demand for shark fins remains a major driver of unsustainable fishing in some markets.

1401 - Some brands commit to selling seafood only from sustainable sources that do not threaten sharks.

1402 - Education programs in schools build a new generation of ocean stewards who care about sharks.

1403 - The precautionary principle guides fisheries to limit shark take when data are uncertain.

1404 - Satellite tagging helps identify critical migratory corridors that deserve protection.

1405 - The ethics of shark conservation include the welfare of sentient animals and the intrinsic value of biodiversity.

1406 - Some nations use no-take zones that extend into offshore waters to protect migratory sharks.

1407 - International agreements help align policies across borders to conserve sharks.

1408 - The number of shark species listed as threatened on the IUCN Red List has risen in recent decades.

1409 - Data gaps in remote regions hinder precise assessments, underscoring the need for ongoing research.

1410 - Bycatch mitigation is most effective when gear changes, timing, and area closures are combined.

1411 - Some fishers report improvements in catch quality and shark survival after adopting bycatch-reduction strategies.

1412 - Captive breeding of sharks is not a common conservation tool for restoring wild populations.

1413 - Public education campaigns can reduce harmful interactions and improve safety around sharks.

1414 - Protective measures around breeding and nursery areas help juvenile sharks survive to adulthood.

1415 - A strong economic case for conservation emerges when sharks support tourism and sustainable livelihoods.

1416 - Illegal, unreported, and unregulated fishing remains a global challenge for sharks.

1417 - International trade rules help ensure shark products come from legal and sustainable sources.

1418 - Population models help scientists estimate declines and forecast recovery times for sharks.

1419 - Public awareness campaigns can shift consumer demand away from high-risk shark products.

1420 - Fisher training and gear changes help reduce bycatch and improve the survival of released sharks.

1421 - Successful shark conservation requires collaboration among scientists, policymakers, fishers, and local communities.

1422 - Protected-area management must be supported by active enforcement and monitoring.

1423 - Recovery of shark populations can take decades, illustrating the long-term commitment required for conservation.

1424 - Individuals can help sharks by supporting conservation organizations, choosing seafood wisely, and sharing what they learn with others.

SHARK SCIENCE AND TECH: HOW WE STUDY THESE OCEAN ICONS

Shark science blends bold fieldwork with sleek technology. In this chapter, you'll peek behind the scenes at the tools researchers use to tag, track, and study sharks without getting in the water with them.

1425 - Tag-and-track methods are the backbone of modern shark science, letting researchers follow movements without constant boatside observation.

1426 - Pop-up satellite archival tags, or PSATs, detach, float to the surface, and beam back data on location, depth, and temperature.

1427 - Archival tags record time-stamped depth and ambient data, building a detailed picture of a shark's daily routine when paired with listening stations.

1428 - Acoustic telemetry networks rely on underwater

receivers that detect tagged sharks as they pass, creating an ocean-wide puzzle of routes.

1429 - Many tags include depth sensors to reveal how deep different species dive during feeding or travel.

1430 - Some tags also measure water temperature, which researchers use to understand how sharks choose their habitats.

1431 - Drones are used to observe surface behavior noninvasively, often helping estimate abundance and schooling patterns.

1432 - Aerial footage from drones can be used to measure body length estimates by comparing known-size objects in the frame.

1433 - Researchers use BRUVs—baited remote underwater video systems—to attract sharks for noncapture observations.

1434 - BRUVs provide species identification, behavior data, and even counts without capturing animals.

1435 - Remotely Operated Vehicles, or ROVs, let scientists videotape and sample sharks in deep or hazardous areas.

1436 - Autonomous Underwater Vehicles, or AUVs, map habitats and track movements using sonar and cameras.

1437 - Electronic tagging has evolved from simple location pings to multi-sensor devices that record depth, temperature, salinity, and sometimes acceleration.

1438 - Data from multiple tag types are integrated into models to predict shark distribution under changing oceans.

1439 - Bite-force studies use force transducers or bite-measuring devices to estimate jaw strength in controlled tests.

1440 - Some bite-force experiments place prey-sized models or safe bite blocks to measure reaction forces of different species.

1441 - Scientists caution that bite-force numbers vary by age, size, and species, so comparisons need context.

1442 - To understand feeding ecology, researchers analyze stomach contents from regurgitated samples or recovered prey remains.

1443 - The subject of bite marks on prey is sometimes used in bite-forensics to identify which shark attacked.

1444 - The unique serrations and tooth shapes of each species aid researchers in identifying sharks from bite marks on prey.

1445 - In situ cameras and lens-filtered imaging capture tooth replacement cycles in captive or observed settings.

1446 - Scientists study tooth wear to infer feeding habits and prey types.

1447 - Dental science helps museum researchers reconstruct historical diets from fossil or specimen teeth.

1448 - Fast, non-invasive population assessments use photogrammetry to estimate body length and growth rates.

1449 - Photogrammetry uses multiple images to calculate size without capturing the animal.

1450 - Some deep-sea sharks use light-producing organs to communicate or confuse predators and prey.

1451 - Researchers study electrosensory capabilities using controlled electrical stimuli in labs and field settings.

1452 - Marine sensors measure changes in ambient electric fields to understand how sharks locate prey.

1453 - Environmental DNA, or eDNA, is collected from water to confirm shark presence in a given area without sighting them.

1454 - eDNA helps monitor rare or cryptic species and supports quick biodiversity assessments.

1455 - Telemetry data are often supplemented with observational data from divers and citizen scientists.

1456 - Ethical guidelines require minimizing stress during tagging; most tagging is temporary and nonlethal.

1457 - Some researchers pilot dive teams with 'shark-safe' equipment like capture nets and protective cages to reduce risk.

1458 - Tag retention can vary; some devices stay on for months, while others shed within weeks.

1459 - Fin-mounted tags may occasionally adapt to the shark's tail movement for better signal stability.

1460 - Researchers are exploring biodegradable tag housings to reduce long-term debris if a device detaches.

1461 - Researchers use barcoded or accelerometer-equipped tags to capture swimming style and vigor.

1462 - Accelerometers reveal whether sharks are cruising, feeding, or resting, by measuring motion.

1463 - Video tags combine a camera with a data logger to record behavior and environmental data.

1464 - Helmet cameras and dorsal-mounted cameras have captured iconic scenes of hunting and schooling.

1465 - Submersible research often uses 'safety lines' to keep operators in secure positions when approaching sharks.

1466 - Tagging and tracking are designed to be non-lethal and minimally invasive.

1467 - Data sharing across institutions accelerates discovery and helps identify global migration patterns.

1468 - Long-term tagging programs reveal that some species migrate across entire ocean basins.

1469 - River sharks, which can live in freshwater, have been tracked to show remarkable adaptability to low-salinity habitats.

1470 - River sharks may undertake long migrations despite living in rivers for much of their lives.

1471 - Tagging has helped identify critical habitats like nursery grounds where juveniles grow in relative safety.

1472 - BRUVs and acoustic arrays can work together to validate predator-prey interactions in a given area.

1473 - Acoustic telemetry requires dense networks of underwater receivers to avoid data gaps.

1474 - The ARGOS satellite system is commonly used to relay tag data from remote ocean regions.

1475 - Some tracking projects use longer-lasting batteries and advanced data storage to extend tag life.

1476 - The choice of tag depends on the species' size, behavior, and the study's questions.

1477 - Wrappers and adhesives used on tagging gear are chosen to minimize tissue irritation and snag risk.

1478 - Habitat mapping uses sonar and side-scan imaging to outline reefs, seafloor features, and potential hunting grounds.

1479 - Ocean models simulate shark movement by combining tracking data with currents and temperature.

1480 - Researchers study diel vertical migration by analyzing depth data from archival tags across day and night.

1481 - Photo-ID techniques, similar to marine mammal studies, help track individual sharks using natural markings.

1482 - Fin clipping is rarely used now due to concerns; noninvasive genotyping from shed skin is preferred.

1483 - Tissue samples collected during tagging enable genetic population studies and diversity estimates.

1484 - Stable isotope analysis reveals shifts in diet across seasons or life stages.

1485 - Researchers use computer vision to automatically identify species from underwater video.

1486 - Deep learning helps sort hours of BRUV footage, speeding up counting and behavior labeling.

1487 - Researchers train and calibrate humans to handle sharks safely during capture-release experiments.

1488 - The 'shark cage' is used to protect divers during close encounters with large individuals.

1489 - Feeding experiments are carefully controlled,

using methods to avoid encouraging predation in the wild.

1490 - The presence of feeding frenzies in the wild is different from scripted feeds in aquariums or films; researchers study the real dynamics.

1491 - Aerial drones help map coastline habitats to identify possible shark hotspots.

1492 - Tagging data sometimes reveal that sharks undertake seasonal long-distance migrations to productive waters.

1493 - Some tags record light levels to estimate latitude and depth.

1494 - Ocean temperature anomalies influence shark distribution and migration speeds; data-logging tags capture this.

1495 - Habitat preferences change with age; juveniles gravitate to nurseries with calm, shallow waters.

1496 - Tagging programs inform conservation by identifying critical corridors to protect.

1497 - Researchers collaborate with fisheries sonar data to cross-check abundance estimates.

1498 - Underwater acoustics can detect not only sharks but other marine life, aiding ecosystem studies.

1499 - The field uses open data repositories to share tracking files with scientists worldwide.

1500 - The public can contribute by reporting sightings or tagging opportunities through citizen science apps.

1501 - Biologging devices collect streams of data that are analyzed with specialized software.

1502 - The life of a tag is a balance between battery life, memory storage, and data resolution.

1503 - Researchers often pre-test tags in aquaria before deploying them in the wild.

1504 - Biopsy darting allows scientists to collect small tissue samples without capturing the entire animal.

1505 - Data from multiple individuals across species reveal common movement patterns and unique strategies.

1506 - Predator-prey dynamics observed with BRUVs illuminate how sharks influence reef ecosystems.

1507 - Light-based tagging is less common but offers discrete tracking options in turbid waters.

1508 - The 'tagger' teams coordinate across time zones to monitor migrations across oceans.

1509 - Sharks can sense electric fields from prey sources even when hidden; the tags help quantify these interactions.

1510 - The field uses careful risk assessment to prevent injury to researchers during fieldwork.

1511 - In addition to tracking, scientists use non-invasive ultrasound to monitor heart rate and physiology in some captive studies.

1512 - Temperature and depth data from tags help researchers model how climate change reshapes shark distributions.

1513 - Researchers deploy underwater camera arrays to document social behaviors and hunting strategies.

1514 - Tagging and tracking have revealed that some species can dive deeper than previously thought.

1515 - Photographs of sharks with tag devices have become iconic for public engagement.

1516 - The use of drones reduces the need to place researchers directly in dangerous situations.

1517 - Conservation messaging stems from science on migration, habitat use, and humane treatment in captivity.

1518 - The field continues to innovate with new materials, sensors, and data tools to unlock a more complete picture of shark life.

www.ingramcontent.com/pod-product-compliance
Lightning Source LLC
Chambersburg PA
CBHW031545220326
41597CB00054B/3459